Descriptive Physical Oceanography

An Introduction

FOURTH ENLARGED EDITION (in SI units)

Some Other Related Pergamon Titles of Interest

BEER   ENVIRONMENTAL OCEANOGRAPHY
An Introduction to the Behaviour of Coastal Waters

THE OPEN UNIVERSITY
The Ocean Basins—Their Structure and Evolution
Seawater—Its Composition, Properties and Behaviour
Ocean Circulation
Waves, Tides & Shallow-water Processes
Ocean Chemistry and Deep-Sea Sediments
Case Studies in Oceanography and Marine Affairs

PARSONS et al
A Manual of Chemical and Biological Methods for Seawater
Analysis

PARSONS et al
Biological Oceanographic Processes, 3rd Edition

POND & PICKARD
Introductory Dynamical Oceanography, 2nd Edition

TCHERNIA
Descriptive Regional Oceanography

WILLIAMS & ELDER
Fluid Physics for Oceanographers & Physicists

Journals

Continental Shelf Research

Deep-Sea Research

Progress in Oceanography

All the above books are available under the terms of the Pergamon text
book inspection copy service. Full details of all Pergamon publications/free
specimen copy of any Pergamon journal available on request from your
nearest Pergamon office.

Descriptive Physical Oceanography

An Introduction

by

George L. Pickard

M.A., D.Phil., D.M.S. (Hon.), F.R.S.C.
Professor Emeritus of Oceanography and Physics
former Director of the Department of Oceanography
University of British Columbia

and

William J. Emery

B.Sc., Ph.D.
Associate Professor
Department of Oceanography
University of British Columbia

FOURTH ENLARGED EDITION (in SI units)

PERGAMON PRESS

OXFORD · NEW YORK · BEIJING · FRANKFURT
SÃO PAULO · SYDNEY · TOKYO · TORONTO

Pergamon Press Offices:

U.K.	Pergamon Press plc, Headington Hill Hall, Oxford OX3 0BW, England
U.S.A.	Pergamon Press, Inc., Maxwell House, Fairview Park, Elmsford, New York 10523, U.S.A.
PEOPLE'S REPUBLIC OF CHINA	Pergamon Press, Room 4037, Qianmen Hotel, Beijing, People's Republic of China
FEDERAL REPUBLIC OF GERMANY	Pergamon Press GmbH, Hammerweg 6, D-6242 Kronberg, Federal Republic of Germany
BRAZIL	Pergamon Editora Ltda, Rua Eça de Queiros, 346, CEP 04011, Paraiso, São Paulo, Brazil
AUSTRALIA	Pergamon Press Australia Pty Ltd., P.O. Box 544, Potts Point, NSW 2011, Australia
JAPAN	Pergamon Press, 8th Floor, Matsuoka Central Building, 1-7-1 Nishishinjuku, Shinjuku-ku, Tokyo 160, Japan
CANADA	Pergamon Press Canada Ltd., Suite 271, 253 College Street, Toronto, Ontario M5T 1R5, Canada

First edition 1964
Reprinted 1966, 1968, 1970
Second edition 1975
Third edition 1979
Fourth enlarged edition 1982
Reprinted 1984, 1985, 1986, 1988, 1989

Library of Congress Cataloging in Publication Data
Pickard, George L.
Descriptive physical oceanography.
Bibliography: p.
Includes index.
1. Oceanography.   I. Emery, William J.   II. Title.
GC150.5.P52   1982   551.46   81-20967   AACR2

British Library Cataloguing in Publication Data
Pickard, George L.
Descriptive physical oceanography:
an introduction—4th enlarged ed
(in SI units)—(Pergamon international library)
1. Oceanography
I. Title   II. Emery, William J.
551.46′01   GC11.2
ISBN 0-08-026280-5 hard cover
ISBN 0-08-026279-1 flexicover

Printed in the United States of America

Preface to the Fourth Edition

THE development of interest in oceanography in recent years has led to an increased demand from students for information on the subject. The texts available hitherto have been either more elementary in treatment or more comprehensive and extensive than may be desirable for an introduction to the subject for the undergraduate. The present text is an attempt to supply information on the Synoptic or Descriptive aspects of Physical Oceanography at a level suitable as an introduction for graduate students, for undergraduates in the sciences and possibly for senior school students who wish to learn something of the aims and achievements in this field of scientific study.

The main object in preparing this fourth edition, in response to suggestions from many users, has been to enlarge significantly the illustrative material with some forty-five additional figures, the original ones having been redrawn, and updated where necessary. The text has also been brought up to date and a selection of references to the original literature has been added. In these tasks, the original author has been fortunate to gain the collaboration of his colleague, Dr. William J. Emery, who has joined him as co-author. A section numbering system has been introduced and there are some minor rearrangements of material. In particular, the discussion of the optical properties of seawater has been moved to Chapter 3 and upwelling to Chapter 8. An elementary discussion of the geostrophic method has been introduced to permit the qualitative deduction of currents from the density distribution.

The International System of Units (SI units) is used as a basis, since this has been recommended by the International Association for the Physical Sciences of the Ocean. For brevity, the term "sverdrup" (Sv), which is not an SI unit, will still be used for volume transports; dissolved oxygen will be expressed in millilitres per litre (mL/L) as most of the existing literature uses this unit, and dynamic metres will be used in the geostrophic method discussion for the same reason. As it may be some time before all oceanographers adopt the SI and to assist in relating to the wealth of literature using the previous mixed system of units, an Appendix describing the two systems is included.

In presenting the synoptic approach it must be emphasized that this represents only one aspect of physical oceanography. The other, and complementary, one is the dynamical approach through the laws of mechanics. This is described in other texts such as those by Knauss, McLellan, Pond and Pickard, Sverdrup *et al.* or Von Arx listed in the Suggestions for Further Reading. The student who requires a full introduction to physical oceanography must study both aspects.

This text is intended to be introductory to the subject. For the student in physics and mathematics it should serve to present the main aspects of the field before he or she proceeds to the more advanced texts and original literature making free use of mathematical methods. For the student in the biological sciences it may provide sufficient information on descriptive physical oceanography to supply the necessary background for studies of the fauna and flora of the sea. The text by Tchernia provides a more detailed coverage of the oceans by regions.

If the reader concludes the text with a feeling that our knowledge of the sea is incomplete at present, one of our objectives will have been achieved. This was to indicate to the student that there is still much to be learned of the ocean and that if he or she is interested in observing the marine world and interpreting it there are still many opportunities to do so.

The Bibliography at the end of the book is in two sections. The "Suggestions for Further Reading" lists a number of texts which would be helpful to a student wishing to read further in descriptive oceanography together with some sources of tables of use to physical oceanographers and a list of some journals which contain articles in the field of oceanography. The "References to Journal and Review Articles" section provides the specific references made by author and date in the body of the text for the convenience of the student who wishes to examine these items in more detail. Suggestions are also offered for more extensive lists of references.

The text is based on a course presented by the authors and colleagues for thirty years at the University of British Columbia to introduce undergraduate and graduate students to physical oceanography. It owes much to the more comprehensive text *The Oceans* by Sverdrup, Johnson and Fleming, and G. L. Pickard wishes to acknowledge this and also the stimulation received during a year at the Scripps Institution of Oceanography. He is particularly indebted to Dr. J. P. Tully of the former Pacific Oceanographic Group for initiating him into oceanography and for encouragement since, and to Dr. R. W. Burling and others for reading the original manuscript and offering constructive comments on that and on subsequent editions of the text. W. J. Emery would like to acknowledge the guidance of Dr. Klaus Wyrtki in introducing him to physical oceanography and providing opportunities for study. Dr. Aas, of the University of Olso, offered some helpful suggestions on the optical aspects and these have been incorporated.

Finally, it will be realized that although the authors have personal experience of some aspects of physical oceanography and of some regions, they have relied very much on the results and interpretations of others in order to present an adequate coverage of the subject. They therefore gratefully acknowledge their indebtedness to the many oceanographers whose works they have consulted in texts and journals in assembling the material for this book.

Acknowledgements

PERMISSION from authors and publishers to reproduce or adapt the following material is acknowledged with thanks:

Fig. 4.6   Data from the Pacific Oceanographic Group, Fisheries Research Board of Canada.

Fig. 4.7   Adapted from Tchernia, 1980, Fig. 7.10, Pergamon Press.

Fig. 5.7 b, c, d, e, f   Adapted from Wyrtki, 1965, Figs. 1–5, *Journal of Geophysical Research*, American Geophysical Union.

Fig. 5.9   From Stommel, 1981, unpublished manuscript.

Fig. 6.5   From Wyrtki, 1974, University of Hawaii.

Fig. 6.6   From Stommel, Niiler and Anati, 1978, Fig. 2, *Journal of Marine Research*, Sears Foundation for Marine Research, New Haven, Connecticut.

Figs. 6.11 and 6.12   Adapted from Figs. 2.2 and 2.9 in L. V. Worthington – The water masses of the World Ocean: some results of a fine-scale census, Chap. 2, pp. 42–69 in *Evolution of Physical Oceanography*, B. A. Warren and C. Wunsch (Eds.) by permission of The MIT Press, Cambridge, Massachusetts. Copyright © 1981 by the Massachusetts Institute of Technology.

Fig. 6.13   From Pickard, 1977, Fig. 2.3, Australian Institute of Marine Science.

Fig. 7.1   From Deacon, G.E.R., 1937, The hydrology of the Southern Ocean, *Discovery Reports*, **15,** 1–123, Cambridge University Press.

Fig. 7.3   From Sverdrup *et al.*, 1946, Fig. 164, Prentice Hall Inc.

Fig. 7.7   From Parker, 1971, Fig. 1, *Deep-Sea Research*, Pergamon Press.

Fig. 7.8   Adapted from Richardson *et al.*, 1978, Figs, 1b, 4, *Journal of Geophysical Research*, American Geophysical Union.

Fig. 7.11   Adapted from Fugilister, 1960, Figs. 23, 29, Woods Hole Oceanographic Institution.

Figs. 7.13, 7.14   Adapted from Sverdrup *et al.*, 1946, Fig. 209B, Prentice Hall Inc.

Fig. 7.16, 7.32   From Emery and Dewar, 1981, Pergamon Press.

Fig. 7.18   Adapted from Wüst, 1961, Fig. 3, *Journal of Geophysical Research*, American Geophysical Union.

Figs. 7.21, 7.22, 7.23   Adapted from Coachman, L. K., 1962, On the water masses of the Arctic Ocean, Department of Oceanography, University of Washington, Report M. 62–11.

Fig. 7.27   Adapted from Rotschi *et al.*, 1972, Figs. 1h, 3h, 4h, 6h, Office de Recherche Scientifique et Technique Outre-Mer, Paris.

Fig. 7.31   Adapted from Wooster and Reid, 1963, Fig. 8, Wiley-Interscience.

Fig. 7.33   Adapted from Tsuchiya, M., 1962, Upper waters of the inter-tropical Pacific Ocean, Fig. 4, Johns Hopkins Press.

Fig. 7.34   Adapted from Reid, 1965, Figs. 2, 3, 4, 6, Johns Hopkins Press.

Plate 4   Courtesy Aanderaa Instruments Ltd.

Plate 7   Courtesy General Oceanics Inc.

Contents

waters and of their movements. The latter include the major ocean currents which circulate continuously but with fluctuating velocity and position, medium and small scale circulation features, the variable coastal currents, the reversing tidal currents, the rise and fall of the tide, and the waves generated by wind or earthquake. The character of the ocean waters includes those aspects, such as temperature and salt content, which together determine density and hence vertical movement, and also includes other dissolved substances or biological species in so far as they yield information about the currents.

The physical study of the oceans is approached in two ways. In what is called the synoptic or descriptive approach, observations are made of specific features and these are reduced to as simple a statement as possible of the character of the features themselves and of their relations to other features. The dynamical or theoretical approach is to apply the already known laws of physics to the ocean, regarding it as a body acted upon by forces, and to endeavour to solve the resulting mathematical equations to obtain inform-ation on the motions to be expected from the forces acting. In practice there are limitations and difficulties associated with both methods, and our present knowledge of the oceans has been developed by a combination of the synoptic and the dynamical approaches. Ideally, the method is as follows. Preliminary observations give one some idea what features of the ocean require explanation. The basic physical law which is considered to apply to the situation is then used to set up an equation between the forces acting and the motions observed. A solution of this equation, even an approximate one, will give some indication of how the motions may vary in time or space. It may also suggest further observations which may be made to test whether or not the law selected or the features entered into the equation are adequate or not. If not, the theory is modified in the light of the test observations and the procedure of alternate observation of nature and development of the theory is pursued until a satisfactory theory is obtained. The method is typical of scientific research.

Our present knowledge in physical oceanography represents an accumu-lation of data, most of which have been gathered during the past hundred years. The purpose of this book is to summarize some of these data to give an idea of what we now know about the distribution of the physical charac-teristics of the ocean waters and of their circulation. The achievements of the alternate but parallel approach through the laws of mechanics are described in other texts such as those by McLellan, Pond and Pickard or Von Arx.

During its history physical oceanography has gone through several phases. Presumably ever since man started to sail the oceans he has been concerned with ocean currents as they affect the course of his ship. This distinctly practical approach is more a branch of the related field of hydrography, which includes the preparation of navigation charts and of current and tide tables, than of oceanography, but out of it came the study of the currents for the purpose of determining *why* they behave in the way they do as well as *how*.

Many of the earlier navigators, such as Cook and Vancouver, made valuable scientific observations during their voyages in the late 1700s, but it is generally considered that Mathew Fontaine Maury (1855) started the systematic large-scale collection of ocean current data, using ship's navigation logs as his source of information. Many physical data on surface currents and winds were collected, and still are, from this source. The first major expedition designed expressly to study all the scientific aspects of the oceans was that of H.M.S. *Challenger* which circumnavigated the globe from 1872 to 1876. The first large-scale expedition organized primarily to gather physical oceanographic data was the German *Meteor* expedition to study the Atlantic Ocean from 1925 to 1927. Expeditions in increasing numbers in the following years have added to our knowledge of the oceans, both in single ship and in multi-ship operations including the loosely coordinated worldwide International Geophysical Year projects in 1957–58, the International Indian Ocean Expedition in 1962–65, the oceanographic aspects of GATE in 1974 (GARP Atlantic Tropical Experiment where GARP = Global Atmospheric Research Programme) and, in the late 1970s, POLYGON, MODE and POLYMODE in the Atlantic (see Chapter 7), the Coastal Upwelling Ecosystems projects in the Pacific and Atlantic, NORPAX (North Pacific Experiment) and ISOS (International Southern Ocean Study) and many other projects. Nevertheless, there are still areas such as the Arctic, the Southern Ocean and the southern Pacific and Indian Oceans for which we have very limited information on which to base our large-scale, steady-state description. Only in a few selected regions do sufficient data exist to allow study of the significant variations in space and time; most of the world's ocean remains a very sparsely sampled environment.

Some of the earliest theoretical studies of the sea were of the surface tides by Newton (1687) and Laplace (1775), and of waves by Gerstner (1847) and Stokes (1874). Following this, about 1896, some of the Scandinavian meteorologists started to turn their attention to the ocean, since dynamical meteorology and dynamical oceanography have much in common. The present basis for dynamical oceanography owes much to the early work of Bjerknes *et al.* (1933), Ekman (1905, 1953), Helland-Hansen (1934) and others.

In recent years attention has been given to other phases, including the circulation and water properties at the ocean boundaries, along the coasts and in estuaries, and also in the deep and bottom waters of the oceans. The coastal waters are more accessible for observation than the open ocean but show large fluctuations in space and time, which present difficulties for theoretical study. For these reasons, the earlier studies tended to be of the open ocean but more detailed studies in recent years have revealed a hitherto unexpected wealth of detail in the form of eddies and shorter-scale time and space variations in the open ocean. The deep and bottom waters are very difficult to observe: this makes it hard to acquire information to start the theoretical studies, and also to test them.

The plan followed in this book will be to describe briefly the ocean basins and something of their topography as it affects ocean circulation, and then to introduce some of the terminology of physical oceanography. After a brief summary of the properties of fresh-water and sea-water, a general description of the distribution of water characteristics both in the vertical and in the horizontal will be presented to give the reader some feeling for typical conditions. A discussion of the sources of gain or loss of heat and water to the ocean follows and then a description of instruments and of methods for data analysis and presentation. After this there is a description of the water characteristics and of the currents in the individual oceans of the world and in coastal regions as such, and finally a few comments on the present state of our knowledge in descriptive physical oceanography.

A comment on the title of this book should be made before proceeding. Among oceanographers the term "synoptic oceanography" is understood to refer to the method of approach which starts with the observation of data and then continues with the preparation of a concise description, i.e. a "synopsis". (The adjective "synoptic" is also used in the term "synoptic data", meaning data for an area collected as quickly as possible to try to eliminate the effects of variations with time.) However, description and synopsis are only the start. The oceanographer then seeks regularities in the data and interprets the distributions of properties with the object of obtaining information on the circulation. Therefore a more exact title for this aspect would be "Interpretative Oceanography", but unfortunately this term is not in general use. "Synoptic Oceanography" was the first choice for a title and would be clear in its meaning for trained oceanographers. However, this book is not intended for them but for would-be oceanographers or for those who desire an introduction to this aspect of science. In the end, the title "Descriptive Physical Oceanography" was chosen in the hope that this would best indicate the character of its contents to those who were not yet familiar with the field. It is hoped, however, that the reader who completes its study will no longer be a novice but will by then appreciate that "synoptic" oceanography does not stop at description but continues with its main aim, interpretation.

CHAPTER TWO

Ocean Dimensions, Shapes and Bottom Materials

2.1 Dimensions

The oceans are basins in the surface of the solid earth containing salt water. The purpose of this chapter is to introduce some of the nomenclature and to direct attention to features of the basins which have a close connection with the circulation and are of importance to the physical oceanographer. A more detailed description of the geology and geophysics of the ocean basins is given in *Submarine Geology* or in *Geological Oceanography*, both by Shepard, or in the articles by Menard and by Bullard in *Oceanography* listed in the Bibliography at the end of this book.

In order to appreciate the shapes of the oceans and seas it is almost essential to examine them on a globe, since map projections on to flat paper always introduce distortions when large portions of the earth are to be represented. From the oceanographic point of view it is convenient to distinguish the various regions in terms of their oceanographic characteristics, particularly their circulations.

Anticipating the information to be given in later chapters, the major ocean areas will be defined now as the Southern Ocean, the Atlantic Ocean, the Pacific Ocean, the Indian Ocean and the Arctic Sea. The last four are clearly divided from each other by land masses but the divisions between the Southern Ocean and the others to its north are determined only by the characteristics of the ocean waters and by their circulations as will be described in Chapter 7. Then there are smaller bodies of water such as the (European) Mediterranean Sea, the Caribbean Sea, the Sea of Japan, the Bering Sea, etc., which are clearly bounded by land or by island chains. The term "sea" is also used for a portion of an ocean which is not divided off by land but has local distinguishing oceanographic characteristics. Examples are the Norwegian, the Labrador and the Tasman Seas.

Looking at a globe again, it is evident that more of the earth's surface is covered by sea than by land, about 71 % compared with 29 %. Furthermore, the proportion of water to land in the southern hemisphere is much greater

5

(4 : 1) than in the northern hemisphere (1.5 : 1). In area, the Pacific Ocean is about as large as the Atlantic and Indian Oceans combined. If one includes the neighbouring sectors of the Southern Ocean with the three main oceans north of it, then the Pacific Ocean occupies about 46 % of the total world ocean area, the Atlantic Ocean about 23 %, the Indian Ocean about 20 %, and the rest combined about 11 %.

The average depth of the oceans is close to 4000 metres while the seas are generally about 1200 m deep or less. Relative to sea level the oceans are much deeper than the land is high. While only 11 % of the land surface of the earth is more than 2000 m above sea level, 84 % of the sea bottom is more than 2000 m deep. However, the maxima are similar: the height of Mt. Everest is about 8840 m while the maximum depth recorded in the oceans is 11,524 m by H.M.S. *Cook* in the Mindanao Trench in the western Pacific. Figure 2.1 shows

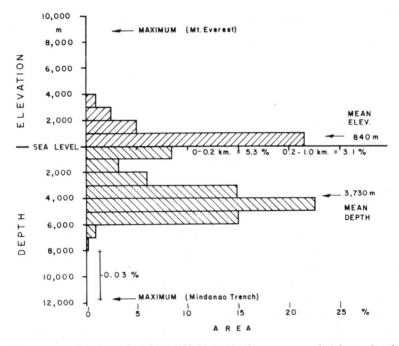

FIG. 2.1.   *Areas of earth's surface above and below sea-level as percentage of total area of earth (in 1000-m intervals).*

the distributions of land elevations and of sea depths relative to sea level in 1000-m intervals as the percentage of the total area of the earth's surface. The land data are from Kossinna (1921) and the ocean data from Menard and Smith (1966).

Although the average depth of the oceans, 4 km, is a considerable distance, it is small compared with the horizontal dimensions of the oceans, which are of the order of 5000 to 15,000 km. An idea of the relative dimensions of the Pacific Ocean may be obtained by stating that they are much the same as a sheet of very thin typing paper. This analogy makes the oceans appear as a very thin skin on the surface of the earth. Relative to the major dimensions of the earth they are thin, but there is a great deal of detail and structure in this thin layer between the sea surface and the bottom of the ocean.

2.2 Sea-floor Dimensions

2.21 Scales

Very often we wish to present some of these details by drawing a vertical cross-section of a part of the oceans. A drawing to true scale would have the relative dimensions of the edge of a sheet of paper and would be either too thin to show details or too long to be convenient. Therefore we usually have to distort our cross-section by making the vertical scale much larger than the horizontal one. For instance, we might use a scale of 1 cm on the paper to represent 100 km horizontally in the sea while depths might be on a scale of 1 cm to represent 100 m, i.e. 0.1 km. In this case the vertical dimensions on our drawing would be magnified by 1000 times compared with the horizontal ones. This gives us room to show the detail but exaggerates the slope of the sea bottom or of lines of constant property (*isopleths*) drawn on the cross-section. It is as well to remind oneself that such slopes are in reality far less than they appear on the cross-section drawings. For instance, a line of constant temperature (*isotherm*) with a real slope of 1 in 100 would be exceptionally steep in the ocean, one of 1 in 1000 very steep and of 1 in 10,000 more usual.

The continents form the major lateral boundaries to the oceans, and the detailed features of the shoreline and of the sea bottom are important in their effects on circulation. Starting from the land, the main divisions recognized are the shore, the continental shelf, the continental slope and rise, and the deep-sea bottom (shown schematically in Fig. 2.2).

2.22 Shore

The *shore* is defined here as part of the land mass close to the sea which has been modified by the action of the sea. It is as well to note in this connection that there is ample evidence to indicate that sea level in the past has varied over a range of about 100 m when glaciers were smaller or larger than they are now. The *beach* is the seaward limit of the shore and extends roughly from the highest to the lowest tide levels. Sandy beaches are often in a state of dynamic equilibrium. That is to say, they may be composed of sand all the time but it may not always be the same sand. This may be continually moving along the

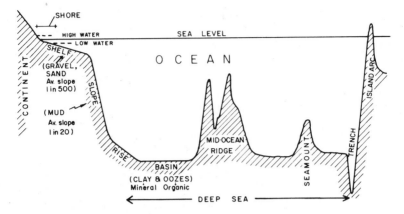

FIG. 2.2. *Schematic section through ocean floor to show principal features.*

shore under the influence of waves and nearshore currents. Evidence for this process can be seen in the way in which sand accumulates against new structures built on the shore, or by the way in which it is removed from a beach when a breakwater is built in such a way as to cut off the supply of sand beyond it. On some beaches, the sand may be removed by currents associated with high waves at one season of the year and replaced by different currents associated with lower waves at another season.

2.23 Continental shelf

The *continental shelf* extends seaward from the shore with an average gradient of 1 in 500. Its outer limit (the "break-in-slope") is set where the gradient increases to about 1 in 20 on the average to form the continental slope down to the deep sea bottom. The shelf has an average width of 65 km. In places it is much narrower that this, while in others, as in the north-eastern Bering Sea or the Arctic shelf off Siberia, it is much wider. The bottom material is dominantly sand, with rock or mud being less common. The division between the shelf and slope is made on the basis of the break-in-slope or "shelf break" which is usually clearly evident when one examines a vertical cross-section of the sea bottom from the shore outward. The average depth at the shelf break is about 130 m. Most of the world's fisheries are located on the continental shelf.

2.24 Continental slope and rise

The *continental slope* averages about 4000 m vertically from the shelf to the deep sea bottom, but in places extends as much as 9000 m vertically, in a relatively short horizontal distance. In general, the continental slope is considerably steeper than the slopes from lowland to highland on land. The material of the slope is predominantly mud, with some rock outcrops. Very typical features of the shelf and slope are the submarine canyons which are of worldwide occurrence. They are valleys in the slope, either V-shaped or with vertical sides, and are usually found off coasts with rivers and never off desert areas. The lower part of the slope, where it grades into the deep-sea bottom is referred to as the *continental rise*.

2.25 Deep-sea bottom and sounding

From the bottom of the continental slope the gradient decreases along the continental rise to the *deep-sea bottom*, the last and most extensive area. Depths of 3000 to 6000 m are found over 74 % of the ocean basins with 1 % being deeper. Perhaps the most characteristic aspect of the deep-sea bottom is the variety of its topography. Before any significant deep ocean soundings were available the sea bottom was regarded as uniformly smooth. When detailed sounding started in connection with cable laying, it became clear that this was not the case and there was a swing to regarding the sea bottom as predominantly rugged. Neither view is exclusively correct, for we know now that there are mountains, valleys and plains on the ocean bottom just as on land. The characteristic features are, as on land, either basically long and narrow (welts and furrows) or of roughly equal lateral extent (swells and basins). The *Mid-Ocean Ridge* is the most extensive feature of the earth's topography. Starting south of Greenland it extends along the middle of the Atlantic from north to south and then through the Indian and Pacific Oceans. In the Atlantic it separates the bottom waters, as can be seen from their very different properties east and west of the ridges (Fig. 7.11). However, there are narrow gaps in this ridge at some of the "fracture zones". These are roughly vertical planes perpendicular to the ridge and on either side of which the crust has moved in opposite directions perpendicular to the ridge. This leaves a gap in the ridge through which water below the ridge top may leak from one side of the ridge to the other. One such is the Romanche Gap through the Mid-Atlantic Ridge close to the equator.

Individual mountains ("seamounts") are widely distributed in the oceans. Some project above the surface to form islands, while the tops of others are below the surface.

In some of the large basins the sea floor is very smooth, possibly more so than the plains areas on land. Stretches of the abyssal plain in the western North Atlantic have been measured to be smooth within 2 m, the present limit of sounding accuracy, over distances of 100 km.

The deepest parts of the oceans are in the trenches. The majority of the deep ones are in the Pacific, including the Aleutian, Kurile, Philippine and Mariana

Trenches, with a few in other oceans such as the Puerto Rico and the South Sandwich Trenches in the Atlantic and the Sunda Trench in the Indian Ocean. In dimensions they are narrow relative to their length and have depths to 10,000 m. They are often shaped like an arc of a circle in plan form with an arc of islands on one side. Because of this they are asymmetrical in cross-section, the island side extending as much as 10,000 m from the trench bottom to the sea surface while the other side is only half as high as it terminates at the ocean depth of about 5000 m.

Our present knowledge of the shape of the ocean floor results from an accumulation of sounding measurements, most of which have been made within the last 60 years. The early measurements were made by lowering a weight on a measured line until the weight touched bottom. This method was slow; in deep water it was uncertain because it was difficult to tell when the weight touched the bottom, and to be certain that the line was vertical. Since 1920 most depth measurements have been made with *echo-sounders* which measure the time taken for a pulse of sound to travel from the ship to the bottom and be reflected back to the ship. One half of this time is multiplied by the average speed of sound in the sea-water under the ship to give the depth. With present-day equipment, the time can be measured very accurately and the main uncertainty over a flat bottom is in the value used for the speed of sound. This varies with water temperature and salinity and if these are not measured at the time of sounding an average value must be used. This introduces a possibility of error. Over trenches or places where rapid changes with depth occur there may also be some uncertainty about whether the echo comes from directly under the ship (to give the true depth) or from one side (to give too small a value).

2.26 Sills

One other term used frequently in reference to bottom topography is *sill*. This refers to a ridge, above the average bottom level in a region, which separates one basin from another or, in the case of a fjord (Chapter 8), separates the landward basin from the sea outside. The *sill depth* is the maximum depth from the sea surface to the top of the ridge, i.e. the maximum depth at which direct flow across the sill is possible.

2.3 Bottom Material

On the continental shelf and slope, most of the bottom material comes directly from the land, either brought down by rivers or blown by the wind. The material of the deep-sea bottom is often more finely divided than that on the shelf or slope. Much of it is pelagic in character, i.e. it has been formed in the sea itself. The two major deep ocean sediments are inorganic "red" clay and the organic "oozes". The former has less than 30 % of organic material and is mainly mineral in content. It consists of fine material from the land (which may have travelled great distances before finally settling out on to the bottom), volcanic material, and meteoric remains. The oozes are over 30 % organic and originate from the remains of living organisms (plankton). The calcareous oozes have a high percentage of calcium carbonate from the shells of animal plankton, while the siliceous oozes have a high proportion of silica from the shells of planktonic plants and animals. The siliceous oozes are found mainly in the Southern Ocean and in the equatorial Pacific. In both of these regions, the distribution is clearly related to the water flow above it—around Antarctica in the first case and parallel to the equator in the second.

Except when turbidity currents (mud slides down the slope) deposit their loads on the ocean bed, the average rate of deposition of the sediments is from 0.1 to 10 mm per 1000 years, and much information on the past history of the oceans is stored up in them. Samples of bottom material are obtained with a "corer" which is a steel pipe 2 to 30 m long which is lowered vertically and forced to penetrate into the sediments by the heavy weight at its upper end. The "core" of sediment retained in the pipe may represent material deposited over a period from 100,000 to 10 million years per metre of length. Sometimes the material is layered, indicating stages of sedimentation of different materials. In some places, layers of volcanic ash can be related to historical records of eruptions; in others, organisms characteristic of cold or of warm waters are found in different layers and suggest changes in temperature of the overlying water during the period represented by the core. In some places gradations from coarse to fine sediments in the upward direction suggest the occurrence of turbidity currents bringing material to the region with the coarser material settling out first and the finer later.

The physical oceanographer looks at the sediments for the information which they give him on the movement of the water above and on its temperature history. This information may be of movements in the past. It may also give some idea of the mean flow, averaging out small fluctuations. The distribution of material around Antarctica is an example of this. The surface of the ocean bottom also reveals information. Photographs of the deep-sea bottom have been obtained in recent years and some of them show ripples such as one sees on a sand beach after the tide has gone out. Such ripples are only found on the beach where the water speed is quite considerable, such as in the backwash from waves. We conclude from the ripples on the deep-ocean bottom that currents of similar speed occur there. This discovery helped to dispel the earlier notion that all deep-sea currents are very slow.

CHAPTER THREE

Physical Properties of Sea-Water

3.1 Vocabulary

In preparation for discussing the physical properties of sea-water in this chapter, and in anticipation of discussions of measurement techniques in Chapter 6, several words will be defined. They are:

Determination: the actual direct measurement of a variable, e.g. the length of a piece of wood with a ruler,

Estimation: a value for one variable derived from the determination of one or more others, e.g. the estimation of salinity from the determination of chlorinity or conductivity.

Accuracy: the difference between a result obtained and the true value.

Precision: the difference between one result and the mean of several obtained by the same method, i.e. reproducibility (includes random errors only).

Systematic error: one which results from a basic (but unrealized) fault in the method and which causes values to be consistently different from the true value (cannot be detected by statistical analysis of values obtained and affects Accuracy).

Random error: one which results from basic limitations in the method, e.g. the limit to the accuracy with which one can read the level of liquid in a burette. It is possible to determine a value for this type of error by statistical analysis of a sufficient number of measurements; it affects Precision. Truly random errors are identified by their Gaussian distribution.

3.2 Properties of Pure Water

Many of the unique characteristics of the ocean can be ascribed to the nature of water itself. Consisting of two negatively charged hydrogen ions and a single positively charged oxygen ion, water is arranged as a polar molecule having positive and negative sides. This molecular polarity leads to water's high dielectric constant (ability to withstand an electric field) and its high

12

solvent ability. Water is able to dissolve more substances than any other fluid. This property explains the abundance of ions in the ocean resulting in its salty character.

The polar nature of the water molecule causes it to form polymer-like chains of up to eight molecules. A certain amount of energy goes into linking these molecules which explains the ocean's ability to absorb heat energy which then may be transported by currents. This aspect of the ocean plays an important, but poorly understood, role in the interaction between sea and atmosphere and the determination of the global climate.

As water is heated, molecular activity increases and thermal expansion occurs. At the same time, added energy is available for the formation of molecular chains whose alignment causes the water to shrink. The combination of these effects results in pure water having a maximum density at 4°C rather than at its freezing point. In sea-water, these molecular effects are overshadowed by the presence of salt which affects the density as described in Section 3.52.

Another important effect of the chainlike molecular structure is the very high surface tension of water. In the ocean, one effect of this can be seen in the formation of surface capillary waves which depend on surface tension for a restoring force. Such capillary waves, despite their small size, are considered to play an important part in determining the friction between wind and water which is responsible for the generation of larger waves and for the major circulations of the upper layers of the oceans.

3.3 Salinity and Conductivity

Sea-water is a complicated solution and contains the majority of the known elements. Some of the more abundant components are chlorine ion 55.0 % of the total dissolved material, sulphate ion 7.7 %, sodium ion 30.6 %, magnesium ion 3.7 % and potassium ion 1.1 %. A significant feature of sea-water is that while the total concentration of dissolved salts varies from place to place, the ratios of the more abundant components remain almost constant. This may be taken as evidence that over geologic time the oceans have become well mixed, i.e. while there are well-marked circulations within each ocean, water must also circulate between the oceans. At the same time there are significant differences in total concentration of the dissolved salts from place to place and at different depths. This indicates that processes must be continually in action to concentrate or dilute sea-water in specific localities; these processes are features of the sea which oceanographers wish to understand.

The total amount of dissolved material in sea-water is termed the *salinity* and has been defined as "the total amount of solid materials in grams contained in one kilogram of sea-water when all the carbonate has been converted to oxide, the bromine and iodine replaced by chlorine and all organic matter completely oxidised". For example, the average salinity of ocean water is about 35 grams of salts per kilogram of sea water (g/kg), usually written as "S = 35°/$_{oo}$" or as "S = 35 ppt" and read at "thirty-five parts per thousand". The direct determination of salinity by chemical analysis or by evaporating sea-water to dryness is too difficult to carry out routinely. The method which was used from the beginning of the century until recently was to determine the amount of chlorine ion (plus the chlorine equivalent of the bromine and iodine), called *chlorinity*, by titration with silver nitrate and then to scale up to the salinity by a relation based on the measured ratio of chlorinity to total dissolved substances (e.g. see Wallace (1974) or Wilson (1975) for a full account). Later, the current definition of salinity was established as "the mass of silver required to precipitate completely the halogens in 0.328 523 4 kg of the sea-water sample". The relation between salinity and chlorinity was redetermined in the early 1960s and since then has been taken as:

$$\text{Salinity} = 1.806\,55 \times \text{Chlorinity}.$$

This is now referred to as the *absolute salinity*, symbol S_A.

However, this definition of salinity has now been replaced by one (called *practical salinity*, symbol S) based on the *electrical conductivity* of sea-water, because almost all salinity estimations are now made by determining this quantity which depends on salinity and temperature (see Lewis and Perkin, 1978; Lewis and Fofonoff, 1979). Because the dependence of electrical conductivity on temperature is significant this has either to be controlled very accurately by a thermostat or measured and corrected for during the measurement. The real advances which permitted the change from chemical titration to the electrical salinometer were refinements in the electrical circuits to permit accurate compensation for temperature. Then the inductive (electrodeless) method (Chapter 6) was developed to avoid problems with electrochemical effects at the electrodes in a conductance cell but recently techniques to eliminate the effects of electrochemical changes have resulted in a return to the electrode-type conductance cell for both laboratory and *in situ* salinometers. The precision of the titration method in routine use is about $\pm 0.02°/_{oo}$ in salinity while by electrical conductivity it is about ± 0.003 in salinity.

The present position with regard to our knowledge of the relations between the chlorinity, the density and the electrical conductivity of sea-water should be made clear. In 1884, Dittmar reported the results of his chemical analysis of seventy-seven samples of sea-water collected from around the world by the *Challenger* Expedition. These results supported the belief, expressed earlier by Forchhammer (1865), that the ratios, one to another, of the concentrations of the major ions in sea-water are subject to only slight variations. Some subsequent writers have rephrased this to imply that the ratios are exactly constant or have acted as though this were the case. That is to say they have assumed that there is an exact and constant relationship between the chlorinity, density and electrical conductivity of sea-water from any place in any ocean and from any depth. In a critical review of the subject, Carritt and Carpenter (1958) pointed out that this rephrasing is incorrect, and that Dittmar's results themselves indicate the possibility of small variations. Such variations may be of significance, particularly in the study of deep water where the differences in properties are small. After Carritt and Carpenter's review, discussions of these matters made it clear that the time had come for the relations to be reinvestigated, both because of the advances made in analytical techniques since Dittmar's time and because of the improvements made in the techniques for the measurement of electrical conductivity of sea-water and in the greater reproducibility of conductivity measurements compared with titration methods.

An extensive series of measurements of the chlorinity, density relative to pure water, and conductivity of sea-water was carried out under the direction of the late R. A. Cox at the National Institute of Oceanography (now the Institute of Oceanographic Sciences) in England on samples from widely distributed locations in the world oceans. The study revealed that the ionic composition of sea-water does show small variations from place to place and from the surface to deep water. The reasons for these variations in ionic composition are not known. It was found that the relationship between density and conductivity was a little closer than between density and chlorinity (Cox et al., 1970). This is interpreted to mean that the chlorinity, i.e. the chemical composition, shows significant differences from place to place and with depth in the oceans. One of the factors used in arriving at this interpretation is that solutions of the major constituents of sea-water of the same concentration have very similar conductivities and densities. This means that the proportion of one ion to another may change, i.e. the chemical composition may change, but as long as the total weight of dissolved substances is the same the conductivity and the density will be effectively unchanged.

Since one of the main reasons for determining either the salinity or conductivity of sea-water is to deduce the density (which is inconvenient to measure directly), the conclusion is that it is better to do this from conductivity than from chlorinity. For this reason, and because of the practical advantages in use (including in situ measurement), the electrical conductivity method is generally used today for the estimation of salinity. (What is actually measured is the ratio of the conductivity of the unknown sea-water sample to that of a standard sea-water of known salinity.) It has been demonstrated that, with average observers, conductivity ratio measurements allow density to be estimated with a precision nearly an order of magnitude better than from chlorinity measurements.

On the Practical Salinity Scale 1978 (PSS 78), the *practical salinity* (S) of a sample of sea-water is defined in terms of the ratio K_{15} of the electrical conductivity of the sea-water sample at the temperature of 15°C and pressure of one standard atmosphere to that of a potassium chloride solution in which the mass fraction of KCl is $32.435\ 6 \times 10^{-3}$ at the same temperature and pressure. A concise review of the earlier definitions of salinity and of the development of the PSS 78 by the Joint Panel on Oceanographic Tables and Standards has been given by Lewis (1980) and includes the formulae for the computation of practical salinity from laboratory and from *in situ* conductivity ratio measurements. (The special issue of the *IEEE Journal of Oceanic Engineering*, Vol. OE-5, No. 1, January 1980, is devoted to (rather technical) details of the PSS 78.)

The International Oceanographic Tables published in 1966 jointly by Unesco and the then National Institute of Oceanography contain the first set of tables for estimating salinity from conductivity ratio values. These tables will be replaced soon by tables based on the PSS 78, but the conversion from conductivity ratio to salinity will usually be carried out by computer using the formulae given by Lewis (1980).

The reasons for the considerable effort by many workers in many laboratories to prepare PSS 78 were that the 1966 Tables only went to 10°C, whereas the bulk of the ocean is at lower temperatures, and the effect of pressure on conductivity needed review, both of these factors affecting *in situ* measurements with CTD instruments with which most salinity data are now collected, and a reproducible primary standard (now the KCl solution) was needed to ensure consistency between laboratories. The PSS 78 is considered valid for the ranges from 2 to 42‰, $-2°$ to 35°C and pressures equivalent to depths from 0 to 10,000 m.

3.4 Temperature

An even more important physical characteristic of sea-water is its temperature. Easily measured with thermometers it was one of the first ocean parameters studied. Even today it continues to be easier to observe temperature and its vertical profile than other properties. In most of the mid- and lower-latitude upper ocean (above 500 m), temperature is the primary parameter determining density, and studies of temperature profiles alone have yielded valuable insights into circulation features. Some of the factors determining temperature are described in Chapter 5 and techniques for its measurement are discussed in Chapter 6. Temperature is always expressed on the Celsius scale in oceanography (°C); when referring to temperature *differences* we shall express them in kelvins (K) to avoid confusion with actual temperatures (°C) (see Appendix).

the conductivity of standard seawater (35‰) at 15°C is 42.914 mmols/cm

3.5 Density

3.51 Units for density

The physical oceanographer is particularly interested in the salinity and temperature of sea-water because they are characteristics which help to identify a particular water body (Chapters 6 and 7) and also because, together with pressure, they determine the *density* (ρ) of sea-water. The latter is important because it determines the depth to which a water mass will settle in equilibrium—the least dense on top and the most dense at the bottom. The distribution of density can also be related to the large-scale circulation of the oceans through the geostrophic relationship (see Chapter 6).

Density is expressed physically in kilograms per cubic metre (kg/m^3) and in the open ocean values range from about $1021.00 \, kg/m^3$ (surface) to about $1070.00 \, kg/m^3$ (at 10,000 m depth). As a matter of convenience, it is usual in oceanography to quote only the last four of these digits in the form of a quantity called $\sigma_{s, t, p}$ defined as:

$$\sigma_{s, t, p} = \text{density} - 1000$$

where s = salinity, t = temperature (Celsius) and p = pressure. This is referred to as the *in situ* value. For many applications in descriptive oceanography the pressure effect on density can be ignored and a quantity $\sigma_{s, t, o}$ is used, commonly abbreviated to σ_t (spoken as "sigma-tee"). This is the density difference of the water sample when the total pressure on it has been reduced to atmospheric (i.e. the water pressure $p = 0$) but the salinity and temperature are as *in situ*. The relationship between σ_t, salinity and temperature is a complicated non-linear one and no simple formula has been devised for it. In practice, values of σ_t are obtained from a nomogram or tables which are entered with the appropriate values of salinity and temperature, or from a polynomial expression in salinity and temperature for computer calculations (e.g. see Millero *et al.*, 1980).

3.52 Effects of salinity and temperature on density

Figure 3.1 shows the σ_t values (curved full lines) for the whole range of salinities and temperatures found anywhere in the oceans. Two points to note are that the change of σ_t with a change of salinity is almost uniform over the whole range of salinity and temperature but that the variation with temperature change is distinctly non-uniform. To emphasize this point, Table 3.1 shows on the left the change of σ_t ($\Delta\sigma_t$) for a change (ΔT) of $+ 1K$ in temperature and on the right the value of $\Delta\sigma_t$ for a change (ΔS) of $+ 0.5°/_{oo}$ in salinity. It will be seen that at high temperatures, σ_t varies significantly with ΔT at all salinities but as temperature decreases the rate of variation decreases,

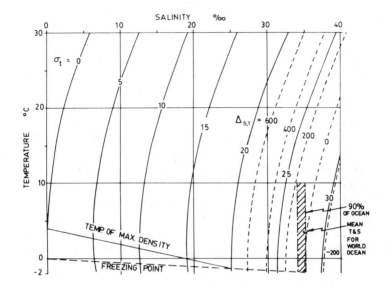

FIG. 3.1.   *Values of σ_t, $\Delta_{S,T}$, temperature of maximum density and freezing point (at atmospheric pressure) for sea-water as functions of temperature and salinity.*

particularly at low salinities (as found at high latitudes or in estuaries). The change of σ_t with ΔS is about the same at all temperatures and salinities. At the same time, it should be noted that the ranges of salinity and temperature found in most of the ocean volume are much less than shown in the whole of Fig. 3.1; the shaded area includes the values for 90 % of the volume of the oceans. More extreme values only occur at the surface. (For convenience, values for σ_t are usually quoted without units because it is generally used for descriptive purposes. Also, as the values for the "density" of sea-water are, in practice, measured relative to that of pure water, the quantities ρ and hence σ_t are

TABLE 3.1

Variation of σ_t ($\Delta\sigma_t$) with variations of temperature (ΔT) and of salinity (ΔS) as functions of temperature and salinity

Salinity ‰	$\Delta\sigma_t$ for $\Delta T = +1C°$			$\Delta\sigma_t$ for $\Delta S = +0.5‰$		
	0	20	40	0	20	40
Temperature (°C)						
30	−0.30	−0.33	−0.34	0.39	0.38	0.38
20	−0.21	−0.24	−0.27	0.40	0.38	0.38
10	−0.09	−0.14	−0.18	0.41	0.39	0.39
0	+0.07	−0.01	−0.17	0.43	0.40	0.40

strictly speaking relative density and relative density difference respectively and are therefore pure numbers without physical dimensions or units. However, in formulae they must be treated as having units of kg/m³.)

In Fig. 3.1 the straight full line represents the temperature at which water has its maximum density, starting at about 4 °C for fresh water, while the dashed line represents the freezing point as a function of salinity. A point to note about freezing is that at low salinities, water which is cooled reaches its maximum density before freezing and sinks while still fluid, the water "overturning" until it all reaches the temperature of maximum density. On further cooling the surface water becomes lighter and finally freezes from the surface down, the deeper water remaining unfrozen. However, at salinities above 24.7‰, as sea-water cools the vertical circulation continues to the freezing point, so that the entire water column must be cooled to this temperature and therefore freezing is delayed.

3.53 Effect of pressure on density

The reason why one can omit the pressure component of $\sigma_{s, t, p}$ is because in descriptive oceanography one is usually comparing water masses at the same depth, i.e. the same pressure, or over the same range of depth. However, it should be noted that the effect of pressure on density is not negligible, i.e. water is not completely incompressible. For instance, a sample of water of salinity 35‰ and temperature 0 °C would have a $\sigma_{s, t, o}$ at the surface (i.e. σ_t) of 28.13 but at a depth of 4000 m its $\sigma_{s, t, p}$ would be increased by compression to 48.49.

Associated with the change of density, i.e. volume, with pressure is a change of temperature if the water does not exchange heat with its surroundings (adiabatic change). For instance, if water of salinity 35‰ and temperature 5.00 °C were lowered adiabatically to a depth of 4000 m its temperature would increase to 5.45 °C due to compression. Conversely, if its temperature were 5.00 °C at 4000 m depth and it were raised adiabatically to the surface it would cool to 4.56 °C due to expansion. This effect has to be considered when water is changing depth significantly. In the example above the temperature of 5.00 °C at 4000 m depth is called the "*in situ*" temperature (T) while 4.56 °C is called the "potential" temperature (θ). The use of the potential temperature is discussed with an example in Chapter 4. The density of the water sample appropriate to its salinity and potential temperature is called its "potential" density and there is a corresponding "*sigma–theta*" (σ_θ).

3.54 Specific volume and anomaly

The *specific volume* (α) is the reciprocal of density, with units m³/kg, and for some purposes is more useful. One application is in the calculation of currents from the distribution of mass by the geostrophic method (Chapter 6). The specific volume *in situ* is written as $\alpha_{s, t, p}$; for convenience a *specific volume anomaly* (δ) is defined as:

$$\delta = \alpha_{s, t, p} - \alpha_{35, o, p}.$$

The last quantity is the specific volume of an arbitrary standard sea-water of salinity $35\,^{0}/_{00}$ and temperature $0\,^\circ C$ at the pressure p. This standard was chosen so that δ is usually positive. Again the connection between α or δ and salinity, temperature and pressure is complicated. The values of $\alpha_{35, o, p}$ are tabulated for all practical values of p (e.g. Sverdrup *et al.*, 1946, p. 1053). For δ, Bjerknes and Sandström (1910) examined the experimental values as functions of salinity, temperature and pressure to show that δ could be broken down into components as:

$$\delta = \delta_s + \delta_t + \delta_{s, t} + \delta_{s, p} + \delta_{t, p} + \delta_{s, t, p}$$

where δ_s represents the main effect of salinity, δ_t represents that of temperature, $\delta_{s, t}$ represents the interacting effects of salinity and temperature, etc. In practice, the last term is small enough to be ignored, while the two pressure terms, $\delta_{s, p}$ and $\delta_{t, p}$ are smaller than the first three terms. In fact, Montgomery and Wooster (1954) pointed out that in the actual oceans the sum of the first three terms $\delta_s + \delta_t + \delta_{s, t} = \Delta_{S,T}$ is adequate in most practical cases to describe the specific volume of water masses. They proposed calling this term, $\Delta_{S,T}$, the *thermosteric anomaly* and in recent years it has come to be used frequently in place of σ_t to describe the density of ocean waters, at least for the upper layers. The physical units for α, the δ's and $\Delta_{S,T}$ are m^3/kg. Numerical values for $\Delta_{S,T}$ are 50 to $100 \times 10^{-8}\,m^3/kg$; for $\delta_{s, p}$ or $\delta_{t, p}$ they are usually of the order of only 5 to $15 \times 10^{-8}\,m^3/kg$ per 1000 m depth. (Previously, a value such as $50 \times 10^{-8}\,m^3/kg$ was written as 50 centilitres per tonne (cL/t) to avoid having to write the power of 10.)

Note that although $\Delta_{S,T}$ is the specific volume analogue of σ_t it is not the reciprocal of σ_t. Convenient formulae for calculating equivalent values of σ_t and $\Delta_{S,T}$ in the range $\sigma_t = 23$ to 28 are:

$$\sigma_t = -0.0105\,\Delta_{S,T} + 28.1 \quad \text{(accurate to 0.1 in } \sigma_t\text{)},$$

$$\Delta_{S,T} = -95.1\,\sigma_t \quad\quad + 2675 \quad \text{(accurate to 1 in } \Delta_{S,T}\text{)}.$$

In using these formulae, the power of 10^{-8} in the physical value of $\Delta_{S,T}$ is ignored for calculation, e.g. for $\sigma_t = 25$, the second formula gives $\Delta_{S,T} = 297$, its physical value then being $297 \times 10^{-8}\,m^3/kg.$.

In Fig. 3.1, isopleths (lines of constant value) for $\Delta_{S,T} = -200, 0, 200, 400$ and $600 \times 10^{-8}\,m^3/kg$ are shown as examples. Isopleths of thermosteric anomaly run substantially parallel to those of σ_t but the values change in the opposite direction. The density of a large proportion of the water of the oceans lies between $\sigma_t = 25.5$ to 28.5 which corresponds approximately to $\Delta_{S,T} = 250$ to $-50 \times 10^{-8}\,m^3/kg$.

3.55 Tables for density and specific volume anomaly

The tables from which values of σ_t or $\Delta_{S,T}$ are obtained are based on laboratory determinations of relative density at different salinities and temperatures. The classical determinations based on measurements by Forch, Jacobsen, Knudsen and Sörensen and presented in the Hydrographical Tables (M. Knudsen, Ed., 1902) give the relative density of sea-water, referred to pure water, as σ_t as a function of chlorinity, salinity and temperature. In 1970 Cox, McCartney and Culkin reported the results of a new investigation of the relations between salinity and density of sea-water relative the pure water. The new results indicated that the values for σ_0 (at T = 0 °C) in "Knudsen's Tables" are low by about 0.01 on the average in the salinity range from 15 to 40°/$_{oo}$ and up to 0.06 at lower salinities and temperatures. In response to a recommendation by the Unesco Joint Panel on Oceanographic Tables and Standards in 1978 a new equation of state was formulated based on all available precision data (Millero et al., 1980) to relate values of α, T, S and p over the ranges of property vales from −2 to 40 °C and 0 to 40°/$_{oo}$ and pressures from 0 to 10^5 kPa (equivalent to 0 to 10,000 m depth). It is considered accurate to 5 × 10^{-6} over oceanic ranges of properties and depths.

For some purposes the possible errors in Knudsen's Tables are not serious. For instance, in the geostrophic method for calculating currents, the oceanographer uses differences between σ_t or $\Delta_{S,T}$ values, not absolute values, and if the salinity and temperature values are not very different, the deficiencies in Knudsen's Tables essentially cancel out and become a minor source of error compared with other sources. This also justifies the oceanographic practice of quoting values of σ_t to four significant figures even though there may be some doubt about the absolute value of the fourth figure.

For plotting σ_t curves, a convenient reference is "Tables for Sigma-T" by Fleming (1939) which give values of salinity and temperature for whole number and fractional values of σ_t from 18 to 30.

The presently available data on density of sea-water are, as mentioned, for the relative density (to pure water). Determinations of the absolute density are needed but this is a very expensive and time-consuming task and only tentative plans are in hand.

One reason why salinity and temperature, and hence density, are important identifying properties of sea-water is because they are "conservative properties" away from the surface. That is to say, below the surface there are no significant processes by which either quantity is changed except by mixing. This absence of in situ sources and sinks means that the spreading of water masses in the ocean can be traced from their origin at the sea surface by their characteristic temperature/salinity values as described in Chapter 6. Near the surface, evaporation or precipitation may change salinity, while many surface heat-transfer processes may change the temperature, as discussed in Chapter 5.

3.6 Other Characteristic Properties

Other characteristics of sea-water which are of help in identifying specific water masses are the dissolved oxygen content, concentration of nutrients (phosphate, nitrate, silicate, etc., ions), plankton, silt, optical characteristics, etc. These, however, have to be used with care because they are not conservative. Biological processes may change the concentration of oxygen or nutrients without any movement of the water mass, silt may settle out (which may also change the optical properties), etc. These other constituents generally occur in such small concentrations that their variations do not significantly affect the density nor do they affect the relations between chlorinity, salinity and conductivity.

The specific heat of sea-water is an important characteristic of which the classical determinations were reported by Thoulet and Chevallier (1889). In 1959 Cox and Smith of the National Institute of Oceanography reported new measurements estimated to be accurate to 0.05 % which gave values 1 to 2 % higher than the old ones. A further study (Millero, Perron and Desnoyers, 1973) gave values in close agreement with those of Cox and Smith. Some of the other thermal properties of sea-water and of sea-ice are described in Chapter 7 in "Ice in the Sea".

3.7 Sound in the Sea

In the atmosphere man receives much of his information about the material world by means of wave energy, either electromagnetic (light) or mechanical (sound). In the atmosphere, light in the visible part of the spectrum is attenuated less than sound, but in the sea the reverse is the case. In clear ocean water, sunlight may be detectable (with instruments) down to 1000 m but the range at which man can see details of objects in the sea is rarely more than 50 m and usually less. Being denied the use of his eyes in the sea, except for close range when diving, man has made much use of sound waves to obtain information. With echo-sounders the depth to the bottom may be measured up to the maximum in the ocean. With SONAR the distance and direction of a submarine may be determined to ranges of hundreds of metres and of fish to somewhat lesser ranges. Side-scan SONARS have been used to take "pictures" of limited parts of the ocean bottom to determine its structure and to locate shipwrecks. The turbulent structure of the ocean and its effects on the propagation of sound waves make it very difficult to form complete images with them. However, following the lead of internal scanning procedures in modern medicine, oceanographers are working to develop "acoustic tomography" whereby they can use a complex pattern of sound waves to provide a picture of the ocean's interior structure.

The speed of sound (longitudinal waves) in water is given by the relation

$V = \sqrt{(E/\rho)}$ where E is the adiabatic compressibility and ρ is the density. As these quantities depend on temperature, salinity and pressure so does the sound speed (Wilson, 1960). The speed of sound at a salinity of 34.85%$_{oo}$ (deep-water average) and 0°C is 1445 m/s. It increases by approximately 4 m/s per K rise of temperature, by 1.5 m/s per 1%$_{oo}$ increase in salinity and by 18 m/s per 1000 m increase in depth (due to the corresponding increase in pressure).

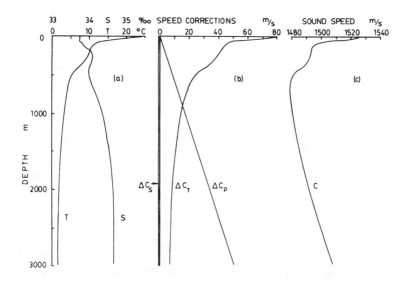

FIG. 3.2. For station in Pacific Ocean at 39° N, 146° W, August 1959: (a) temperature and salinity profiles, (b) corrections to sound speed due to salinity, temperature and pressure, (c) resultant in situ sound-speed profile showing sound-speed minimum.

A consequence of these variations of sound speed with water properties and depth and the typical vertical distributions of these properties is an *in situ* sound speed minimum at depths ranging from near the surface at high latitudes (low temperatures in the upper water) to over 1000 m at low latitudes. Figure 3.2 shows typical temperature and salinity profiles for a mid-latitude station together with the individual effects of property and depth variations on sound speed. The resultant sound speed profile with a minimum centred at about 700 m depth is also shown. Note that the minimum results chiefly from the reduction in sound speed with decreasing temperature combined with the increase in speed as depth (i.e. pressure) increases. Salinity variations have very little effect on sound speed. The wider range of temperature variations in the ocean and their stronger effect on sound speed changes make it possible to study acoustic propagation with vertical temperature profiles alone.

A consequence of the sound-speed minimum is that sound waves tend to be trapped within it. For waves which are directed upward, the increase of speed as they move toward the surface refracts (redirects) them down, while waves directed down tend to be refracted up. The result is a *sound channel* (called the SOFAR channel for SOund Fixing And Ranging) along which sound energy may travel very considerable distances in the sea, a characteristic which aids in the detection of submarines at long ranges and is used for tracking free-drifting subsurface floats as will be described in Chapter 6. Low-frequency sound (hundreds of hertz) may be detected after travelling thousands of kilometres in the SOFAR channel.

As the speed (V), wavelength (λ) and frequency (n) are connected by the wave equation $V = n\lambda$, the wavelength for a frequency of 10 kilohertz is about 14 cm and for a frequency of 100 kHz is about 1.4 cm. Most echo-sounders and sonars operate at frequencies in this range. The speed does not depend on frequency and so sound waves are non-dispersive in the sea.

Since the size of the sound source of an echo-sounder is not much larger than the wavelength, the angular width of the sound beam emitted is large. With a 12-kHz echo-sounder the beam width, in which the energy is no less than one-half of its maximum, will be of the order of 30° to 60°, which makes it difficult to distinguish details of the bottom topography. It is possible to improve the resolution by using higher frequencies to 100 or even 200 kHz but as the absorption of sea-water for sound energy increases as the square of the frequency, this may entail loss of range or depth of water which can be penetrated.

In echo-sounding applications, the sound energy is directed vertically downward and the only effect of changes of water properties is to modify the speed of sound and so introduce a possible error when determining the depth from the formula: depth $= \frac{1}{2}t V$ where t is the time taken for a sound pulse to travel from the ship to the sea bottom and back. In open ocean areas the speed of sound varies with depth over a range of a few tens of metres per second and for echo-sounding purposes an average is used (e.g. 1463 m/s = 4800 ft/s = 800 fathoms per second), with corrections determined by the water properties if accurate depths are needed.

The ratio of the speed of sound in water to that in air (about 4.5 to 1) is large, and so only a small amount of the sound energy starting in one medium will penetrate through the interface into the other, in contrast to the relative efficiency of passage of light energy through the water/air interface (speed ratio only 1 to 1.33). This is the reason why a man standing on the shore is unaware of the noises in the sea, and why it is not possible to speak directly from the air to a diver under water. However, it is possible to design mechanical transducers, such as hydrophones or echo-sounder sources, to pick up sound energy from the water efficiently.

3.8 Light in the Sea

The behaviour of visible light in water is different in degree from that in air, in particular it is absorbed in much shorter distances in the sea than in the atmosphere. We are concerned here with short-wave energy of wavelengths from about 0.4 to 0.8 μm (1 μm = 10^{-6} m), i.e. from the violet to the red of the visible spectrum. When this short-wave energy penetrates into the sea, some is scattered but most is absorbed and causes the temperature of the water to rise. It is the major source of supply of heat to the oceans. The vertical attenuation of the energy is progressive as the radiation passes down through the sea and is different for different wavelengths. The progressive decrease in energy penetrating downward is expressed by an exponential law, $I_z = I_0 \exp(-kz)$, where I_0 is the radiation intensity penetrating through the surface, I_z is the remaining intensity at a depth z metres below the surface and k is the *vertical attenuation coefficient* of the water. The effects of depth and of attenuation coefficient are shown in Table 3.2. The coefficient k depends mainly on the absorption of light in the water and to a lesser extent on scattering.

TABLE 3.2

Amount of light penetrating to specified depths in sea-water as a percentage of that entering through the surface

Depth z (m)	Vertical attenuation coefficient, k (m^{-1})			Clearest ocean water	Turbid coastal water
	0.02	0.2	2		
0	$I_0 = 100\%$	100%	100%	100%	100%
1	$I_z = 98$	82	14	45	18
2	96	67	2	39	8
10	82	14	0	22	0
50	37	0	0	5	0
100	14	0	0	0.5	0

The first three columns after the depth column demonstrate the relative influence of the attenuation coefficient k and of the depth z on energy of a particular wavelength; the last two columns represent practical conditions in the sea and will be explained below. The column of the table for $k = 0.02$ demonstrates transmission of the most penetrating component, blue light, passing through the clearest ocean water. Energy penetrates coastal waters less readily than this because of the extra attenuation due to suspended particulate matter and to dissolved materials. The absorbing materials are probably chiefly organic acids, the inorganic salts do not show significant absorption at visible wavelengths. The columns for $k = 0.2$ and 2 are more indicative of the transmission of blue light through less clear ocean water and through very turbid water respectively.

For sea-water, the coefficient k varies considerably with wavelength. For clear ocean water (Fig. 3.3(a), curve 1) k has its minimum value at about 0.45 μm wavelength so that blue light is attenuated least (penetrates best) while at shorter wavelengths (toward the ultra-violet) and at longer wavelengths (in the red and infra-red) the attenuation is much greater and penetration correspondingly less. The increased attenuation in the ultra-violet is not important to the heat budget of the oceans because the amount of energy reaching sea level at such short wavelengths is small. The increased absorption is more important in and beyond the red end of the spectrum where more energy is present in the sun's radiation. Virtually all of the energy shorter than the visible is absorbed in the top metre of water, while the energy of wavelength 1.5 μm or greater is absorbed in the top one centimetre or less. In Fig. 3.3(b) (full lines) are shown the relative amounts of energy penetrating clear ocean water to 1, 10 and 50 m

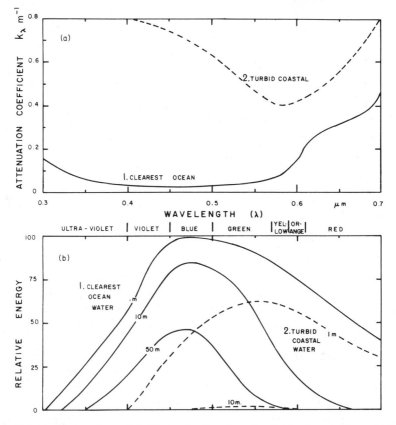

FIG. 3.3. (a) Attenuation coefficient k_λ as a function of wavelength λ for clearest ocean water (full line) and turbid coastal water (dashed line). (b) Relative energy reaching 1, 10 and 50 m depth for clearest ocean water and reaching 1 and 10 m for turbid coastal water.

as a function of wavelength. The maximum penetration is in the blue while penetration by yellow and red is much less.

In more turbid waters, e.g. near the coast, all wavelengths are attenuated more than in clear water, i.e. k is larger as shown by curve 2 in Fig. 3.3(a), and the least attenuation is in the yellow part of the spectrum. The relative penetration of energy for turbid coastal water is shown by the dashed lines in Fig. 3.3(b). The energy penetrating to 1 m is less and to 10 m much less than through clear water, and the maximum penetration is shifted to the yellow. (The energy reaching 50 m in this turbid water is too small to show on the scale of this graph.)

In clear ocean water the superior penetration of blue and green light is evident both visually when SCUBA diving and also in colour photographs taken underwater by natural light. Red or yellow objects appear darker in colour or even black as they are viewed at increasing depths because the light at the red end of the spectrum has been absorbed in the upper layers and little is left to be reflected by the object. Blue or green objects retain their colour to greater depths. In the clear ocean water there is enough light at 50 to 100 m to permit a diver to work, but in turbid coastal waters almost all of the energy may have been absorbed by 10 m depth.

The above remarks have emphasized the penetration of particular wavelengths. The energy from the sun is composed of a range (spectrum) of wavelengths and, as far as the heat budget is concerned (see Chapter 5), the important quantity is the sum total of energy at all wavelengths penetrating into the water. In any particular body of sea-water the total energy penetration is less than that of the most penetrating component. The last two columns of Table 3.2 indicate the range of penetrations found in actual sea-water.

In addition to its significance for the heat budget, the penetration of radiation into the sea is of interest to biologists in connection with the photosynthetic activity of phytoplankton and algae and with the behaviour of zooplankton and fish. Some of the methods for measuring the attenuation or transmission of radiation through the sea are described in Chapter 6.

3.9 Colour of Sea-Water

A number of investigators have considered the reasons for the colour of the sea which ranges from deep blue to green or even greenish-yellow (e.g. Jerlov, 1976). The number of records of sea colour is not great but broadly speaking the deep or indigo blue colour is characteristic of tropical and equatorial seas, particularly where there is little biological production. At higher latitudes, the colour changes through green-blue to green in polar regions. Coastal waters are generally greenish. There are two factors contributing to the blue colour of open ocean waters at low latitudes where there is little particulate matter. In deep water if one looks downward from below the surface, as when snorkelling, the light which one sees is mainly that scattered by the molecules of the water. Because the molecules scatter the short-wave (blue) light much more than the long-wave (red) light the colour seen is selectively blue. In addition, because the red and yellow components of sunlight are rapidly absorbed in the upper few metres, the only light remaining to be scattered from the bulk of the water is the blue light. If one looks at the sea from above the surface, in addition to the blue light scattered from the body of the water one sees some sky light reflected from the surface and the two components add together. If the sky is blue, the sea will still appear deep blue, but if there are clouds the white light reflected from the sea surface will dilute the blue scattered light from the water and the sea will appear less intensely blue. If there are green phytoplankton on the water their chlorophyll content will absorb the blue light and shift the water colour to green. The organic products from plants may also add yellow dyes to the water and these will absorb blue and shift the apparent colour toward the green. This is the situation in the more productive high latitude and coastal waters. In some coastal regions, rivers bring in dissolved organic substances which emphasize the yellowish-green colour. The red colour occurring sporadically in some coastal areas, the so-called "red tide", is caused by blooms of species of phytoplankton of a reddish-brown colour. In other regions, rivers bring in finely divided inorganic material, mud and silt, which may impart their own colour to the water by reflection from the particles. In fjords fed by rivers from glaciers, the surface low-salinity layer may be milky-white from the finely divided "rock flour" produced by abrasion in the glaciers and carried down by the melt water. The material may be kept in suspension by turbulence in the upper layer for a time but when it sinks into the saline water below it flocculates and sinks more rapidly. When diving in such a region one may be able to see only a fraction of a metre in the upper layer but be able to see several metres in the saline water below. The colour of sea-water can be judged most conveniently against the white Secchi disc (Chapter 6) as it is lowered to determine the transparency of the upper water. To judge the colour one may use the "Forel Scale" provided by a set of glass tubes of different shades of blue-coloured water.

CHAPTER FOUR

Typical Distributions of Water Characteristics in the Oceans

4.1 Introduction

4.11 General

In the previous chapter attention was drawn to temperature and salinity as ocean water characteristics. These quantities vary from place to place in the ocean, and from their distribution we can learn a good deal about the average circulation of the waters. In this chapter, some of the typical distributions will be described so that the reader may gain some feeling for them and be able to recognize normal and abnormal distributions.

A salient feature of the distribution of many water characteristics is that they are horizontally stratified or nearly so. In other words, the sea is made up of substantially horizontal layers as far as these characteristics are concerned and horizontal changes are generally much smaller than vertical ones in the same distance. For instance, near the equator the temperature of the water may drop from 25° C at the surface to 5° C at a depth of 1 km, but it may be necessary to go 5000 km north or south from the equator to reach a latitude where the surface temperature has fallen to 5° C. The average vertical temperature gradient (change of temperature per unit distance) in this case is about 5000 times the horizontal one. However, the horizontal variations do exist and therefore the water properties are distributed in three dimensions. This makes it difficult to display them when we are limited to plotting on paper with only two dimensions. We are usually forced to represent a single real three-dimensional distribution by a number of two-dimensional ones, such as vertical and horizontal sections.

4.12 Collection and analysis of data

To appreciate the way in which our understanding of these distributions is built up, it is helpful to describe briefly how the data are collected. The oceanographer goes to sea for a cruise in a research ship (Pl. 1) to "occupy a number of oceanographic stations" (sometimes called "hydrographic stations"). This means that at a number of preselected locations (*stations*) the ship is stopped and the oceanographers measure the water properties from surface to deep water. For each station the measurements are then plotted as "vertical profiles", i.e. graphs of temperature, salinity, etc., against depth. The vertical profiles from stations along a line may then be put together to form a "vertical section" of water properties, or the data from a selected depth may be plotted and contoured to show the horizontal distribution of the property at that depth like a map. These are the synopses of the spatial distributions of the data.

After the field work and the subsequent plotting of the data, the really interesting part starts. The oceanographer sits down with the plots before him to try to determine the reasons why they are as they are. Among the processes which may be acting are horizontal flow, sinking or upwelling, mixing, diffusion, heat flow through the surface, precipitation etc. There is no set routine method that will be suitable for all situations. The oceanographer does exactly what a detective does in trying to solve a crime. Each assembles all the data (clues) that he can and then tries to deduce from them what really happened and what forces (motives) may be relevant. Each may have all the clues needed, or may lack vital ones, or may be looking at the wrong ones. The oceanographer perhaps has one advantage over the detective. He can sometimes go back to make more observations, whereas the criminal may not be so obliging as to keep repeating his crime.

There is one more dimension to be considered. In addition to the spatial variations of properties (vertical and horizontal) there may be variations with time ("temporal" variations). To observe these the oceanographic ship occupies an *anchor station*, i.e. anchors for a day, a week or longer, or else returns to the same station, time after time, to measure the water properties there. An alternative and more usual approach nowadays is to anchor instrumentation such as current meters, temperature sensors, etc., at specific locations. These instruments automatically record for a period of time (sometimes as long as a year) and are then retrieved for subsequent analysis of their data. These are plotted as *time series* in various ways to see if there are daily ("diurnal") or seasonal variations. Generally the diurnal variations are only appreciable to depths of a few metres and seasonal ones to 100 to 300 m. At greater depths variations take place only over periods of weeks, years or possibly centuries.

4.13 General statistics and area descriptions

Before giving descriptions of some of the typical distributions of water properties, the following statistics on ocean water temperatures and salinities are given for orientation:

(a) 75 % of the total volume of the ocean water has properties within the range from 0° to 6° C in temperature and 34 to 35 °/$_{oo}$ in salinity,

(b) 50 % of the total volume of the oceans has properties between 1.3° and 3.8° C and between 34.6 and 34.8 °/$_{oo}$,

(c) the mean temperature of the world ocean is 3.5° C and the mean salinity is 34.7 °/$_{oo}$.

Referring back to Fig. 3.1 it will be seen that the ranges of temperature and salinity found in most of the oceans cover only a very small part of the extreme ranges that may be found. Near the end of Chapter 6 will be given more information on the volumes of ocean water within various ranges of temperature and salinity individually and, what is more important in identifying particular water masses, on the ocean volumes having certain *combinations* of temperature and salinity values.

One feature to be noted about the spatial distribution of water properties is that in the surface and upper waters there is a distinct tendency for the arrangement of some of them to be "zonal", i.e. the value of a property may be much the same across the ocean in the east—west direction but may change rapidly in the north—south direction. As a consequence, when describing ocean water property distributions we frequently wish to refer to the position of an ocean area in terms of its north–south position even when we do not need to specify the latitude explicitly. For such zonal distributions, the adjective "equatorial" refers to the zone near the equator, while the adjective "tropical" refers to zones near the tropics (23½° N or S of the equator). The distinction between "equatorial" and "tropical" should be noted as it is often significant; when the two are to be lumped together the term "low latitude" will be used, in contrast to the "high latitudes" which are near the poles, north and south. "Subtropical" refers to zones on the high latitude side of the tropical zones. The term "polar" is properly applied to oceanography only to the Arctic regions but is often used of the ocean close to Antarctica.

4.2 Density Distribution

4.21 Density at the surface

The distribution of the density of sea-water at the ocean surface can be described roughly by stating that the value of σ_t increases from about 22 near the equator to 26 to 27 at 50° to 60° latitude, and beyond this it decreases slightly (Fig. 4.1).

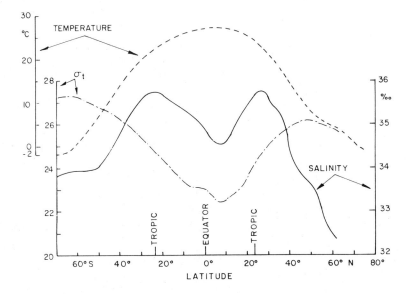

FIG. 4.1.   *Variation with latitude of surface temperature, salinity and density* (σ_t)–*average for all oceans.*

4.22 Subsurface density and the pycnocline

More important, however, is the distribution of density in the vertical direction. A guiding principle here is that the density normally increases as depth increases. This is simply a consequence of the general tendency in Nature for a system to settle down to a state of minimum energy. This is the case in still water when the least dense water is at the surface and the most dense at the bottom. The density in the sea does not increase uniformly with depth, however. In equatorial and tropical regions there is usually a shallow upper layer of nearly uniform density, then a layer where the density increases rapidly with depth, called the *pycnocline*, and below this the deep zone where the density increases more slowly with depth (Fig. 4.2). There is little variation with latitude of the deep water σ_t which is about 27.9. In consequence, in high latitudes where the surface σ_t rises to 27 or more there is a much smaller increase of density with depth than in the low latitudes and the pycnocline is less evident.

4.23 Static stability

The rate of change of density with depth determines the water's *static stability* or unwillingness to be moved vertically. Where the stability is high, vertical movement and vertical mixing are minimized. A measure of stability is

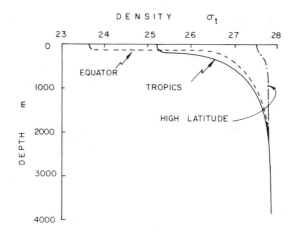

FIG. 4.2. *Typical density/depth profiles for low and high latitudes.*

the value of the quantity $\delta\sigma_\theta/\delta z$ ($\delta\sigma_t/\delta z$ may be used in the upper layers). If its value is positive, the water is stable, if negative it is unstable: i.e. if σ_θ increases with increasing depth the water is stable, if it decreases with increasing depth the water is unstable. Where there is no change of potential density (σ_θ) with depth, i.e. $\delta\sigma_\theta/\delta z = 0$, the water has neutral stability and may be mixed vertically with no effort. In the surface layer (50 to 100 m) slight instability often occurs in mid-latitudes, possibly due to increases in salinity resulting from evaporation. Unstable conditions are uncommon below the surface layer; they may be found near the interface between bodies of water of different density distributions in the process of mixing, e.g. at the northern edge of the Gulf Stream. (A more detailed discussion of stability will be found in Pond and Pickard, *Introductory Dynamic Oceanography.*)

The water in the pycnocline is very stable. That is to say, it takes much more energy to displace a particle of water up or down in the pycnocline than in a region of lesser stability. A result is that turbulence, which causes most of the mixing between different water bodies, is less able to penetrate through this layer than through the less stable water. The pycnocline then, although it is too slight to offer any barrier to the sinking of bodies which are much denser than water, offers a real barrier to the passage of water and water properties in the vertical direction, either up or down. Due to the presence of the pycnocline, the vertical density distribution can be approximately represented as a thin, upper low-density layer of 100 to 500 m thickness lying over the remainder of the ocean with a near-uniform higher density. The so-called "two-layer" ocean is a convenient simplified description often used in analytical and numerical ocean models. Results of these models are expressed in terms of variations in the position of the interface between the two layers which represents the pycnocline. The use of two-layer models is more completely discussed by Pond and Pickard in *Introductory Dynamic Oceanography*.

4.24 Geographic distribution of density

Perhaps the most illuminating graphical presentation of the distribution of density is a north–south vertical section through the ocean, such as Figs. 7.9 and 7.34, showing isopycnals of σ_θ and σ_t respectively. Those in the upper layers tend to be concave upward, showing the increase from equator to pole. Below about 2000 m, however, the total range of values is only from about 27.6 to 27.85 for σ_t or 27.8 to 27.95 for σ_θ.

A further point to bear in mind when considering the ocean circulations later is that there is a strong tendency for flow to be along surfaces of constant potential density. In the upper layers we can regard this as to be along surfaces of constant σ_t. For instance, the processes which give ocean waters their particular properties act almost exclusively at the surface, and one can trace the origin of even the deepest water back to a region of formation at the surface somewhere. Since deep ocean water is of high density this implies that it must have formed at high latitudes because only there is high density water found at the surface. After formation it sinks along constant density surfaces. When the word "sink" is used here it does not necessarily imply that the water goes straight down like a stone. The sinking is often combined with horizontal motion so that the water actually moves in a direction only slightly inclined below the horizontal. The slopes of the constant density surfaces in Figs. 7.9 and 7.34 are exaggerated because of the vertical scale exaggeration used in plotting the data.

In the low- and mid-latitude open ocean, most of the variations of density in the upper 1000 m are due to variations of temperature; at greater depths, salinity variations may play a significant part. The effect of salinity on the deep water is more evident in the Atlantic (Fig. 7.9), where there is an obvious salinity structure, than in the Pacific (Fig. 7.34) where the deep waters are more uniform. Only in certain areas, such as the north-east Pacific and in the polar regions, does the variation in salinity play a part in the upper layers. In coastal waters, fjords and estuaries, salinity is often the controlling factor in determining density at all depths, while the temperature variations are of secondary importance. The information in Table 3.1 is useful in estimating the relative effects of temperature and salinity changes on density in a particular situation.

4.3 Temperature Distribution

4.31 Surface temperature

The distribution of temperature at the surface of the open ocean is approximately zonal, the lines of constant temperature (*isotherms*) running roughly east–west (Figs. 4.3 and 4.4). (The purpose of the projection in this and some subsequent figures is to show the ocean with areas between parallels of latitude being reasonably correct, rather than being grossly exaggerated at high latitudes as they are on Mercator's projection.) Near the coast where the currents are diverted the isotherms may swing more nearly north and south. Also, along the eastern boundaries of the oceans low surface temperatures often occur due to upwelling of subsurface cool water, e.g. along the west coast of North America in summer, which causes the isotherms to trend equatorward. (Upwelling will be discussed in Chapter 8.) The open ocean surface temperature decreases from as high as 28° C just north of the equator to nearly $-2°$ C near ice at high latitudes. Values for the surface temperature from south to north for all oceans averaged together are presented in Fig. 4.1. This distribution with highest values at low latitudes and decreasing at higher latitudes corresponds closely with the input of short-wave radiation as will be discussed in Chapter 5.

4.32 Upper layers and the thermocline

Below the surface the water can usually be divided into three zones in terms of its temperature structure (Fig. 4.5). There is an upper zone of 50 to 200 m depth with temperatures similar to those at the surface, a zone below this extending from 200 to 1000 m in which the temperature decreases rapidly, and a deep zone in which the temperature changes slowly. Typical temperatures at low latitudes would be 20° C at the surface, 8° C at 500 m, 5° C at 1000 m and 2° C at 4000 m.

The depth at which the temperature gradient (rate of decrease of temperature with increase of depth) is a maximum is called the *thermocline*. With actual observations of temperature in the sea it is often difficult to determine this depth precisely because of minor irregularities in the temperature/depth profile and it is easier to pick out a "thermocline zone" as a range of depth over which the temperature gradient is large compared with that above and below. Even for this zone, it is often hard to define precisely the depth limits, particularly the lower limit, and one must accept some degree of approximation in stating the depth limits of the thermocline zone. However, in low and middle latitudes it is clear that there is a thermocline present all the time at depths between 200 and 1000 m. This is referred to as the "main" or "permanent" thermocline. In polar waters there is no permanent thermocline.

The continued existence of the thermocline requires explanation. One might expect that as the upper waters are warmest, heat would be transferred downward, despite the inhibiting effect of the pycnocline, and that the temperature difference between the upper and lower layers would eventually disappear. This does not seem to be taking place. One suggestion is that while downward transfer of heat does occur in mid- and low-latitudes (primarily by diffusion), there is a simultaneous upward movement (advection) of cool

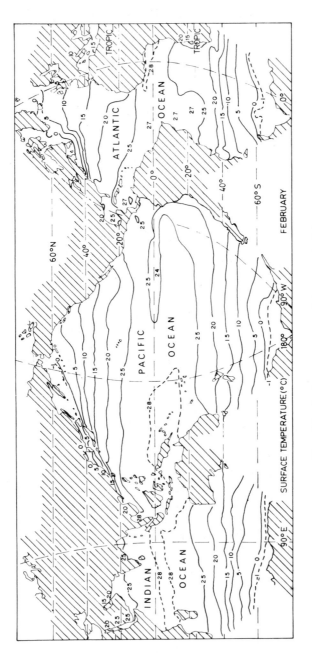

FIG. 4.3.   Surface temperature of the oceans in February.

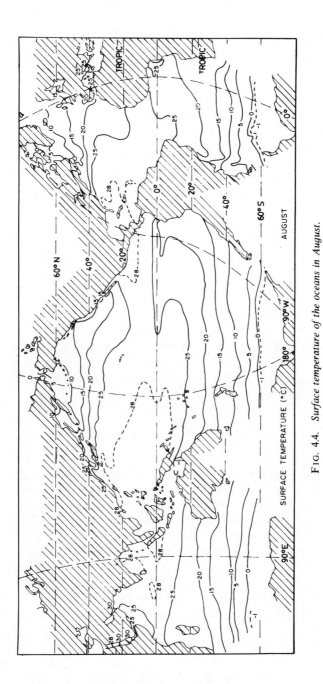

FIG. 4.4. *Surface temperature of the oceans in August.*

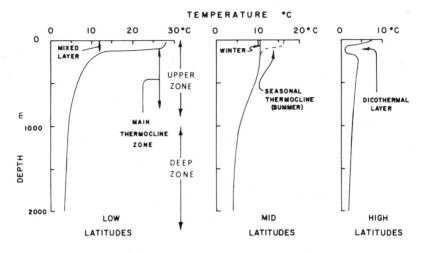

FIG. 4.5.   *Typical mean temperature/depth profiles for the open ocean.*

water from below in these latitudes as a consequence of the sinking of cold
water at high latitudes, as discussed in Chapter 7. It is presumed, then, that the
two processes are in long-term balance and the thermocline is a permanent
feature of much of the ocean's upper layer. This is just a basic mechanism, the
differences in the shape of the thermocline in different regions still require
detailed explanation. It is possible that other mechanisms such as lateral
mixing and flow may play a part.

4.33 Temporal variations of temperature in the upper layer.

The temperature in the upper zone shows seasonal variations, particularly
in middle latitudes. The layer between the surface and a depth of 25 to 200 m is
usually at much the same temperature as the surface water because of mixing
due to wind waves. For this reason it is referred to as the *mixed layer* (Fig. 4.5).
In winter the surface temperature is low, waves are large, and the mixed layer
is deep and may extend to the main thermocline. In summer the surface
temperature rises, the water becomes more stable, and a "seasonal" thermocline often develops in the upper zone (Fig. 4.5). The thermocline zones are of
high stability (they are essentially the pycnocline zones), and for this reason
they separate the waters of the upper from those of the deep zones.

An illustration of the growth and decay of the seasonal thermocline is
shown in Fig. 4.6(a). This figure shows monthly temperature profiles from
March 1956 to January 1957 taken at Ocean Weather Station "P" in the
eastern North Pacific. From March to August the temperature gradually
increases due to absorption of solar energy. A mixed layer from the surface

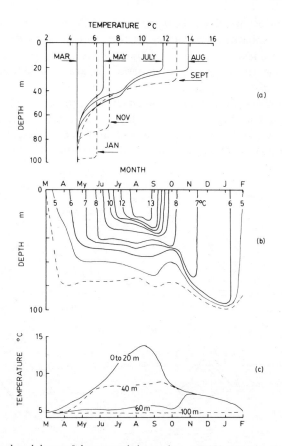

FIG. 4.6. *Growth and decay of the seasonal thermocline at 50° N, 145° W in the eastern North Pacific.*

down to 30 m is evident all the time. After August there is a net loss of heat energy from the sea while continued wind mixing erodes away the seasonal thermocline until the isothermal condition of March is approached again.

The same data may be presented in alternative forms. In Fig. 4.6(b) is given a time-series plot showing the depth of the isotherms during the year. (The original data include the alternate months which were omitted from Fig. 4.6(a) to avoid crowding.) In Fig. 4.6(c) are plotted the temperatures at selected depths, i.e. *isobaths* of temperature. The different forms in which the thermocline appears in these three presentations should be noted. In Fig. 4.6(a) it appears as a maximum gradient region in the temperature/depth profiles. In Fig. 4.6(b) the thermocline appears as a crowding of the isotherms which rises from about 50 m in May to 30 m in August and then descends to 100 m in January. In Fig. 4.6(c) the thermocline appears as a wide separation of the 20-m and 60-m isobaths between May and October, and between the 60-m and 100-m isobaths after that as the thermocline descends.

At high latitudes the surface temperatures are much lower than at lower latitudes, while the deep-water temperatures are little different. In consequence the main thermocline may not be present and only a seasonal thermocline may occur. In high northern latitudes there is often a *dicothermal layer* at 50 to 100 m (Fig. 4.5). This is a layer of cold water, down to −1.6°C, sandwiched between the warmer surface and deeper layers. (Stability is maintained by an increase of salinity with depth through the layer.)

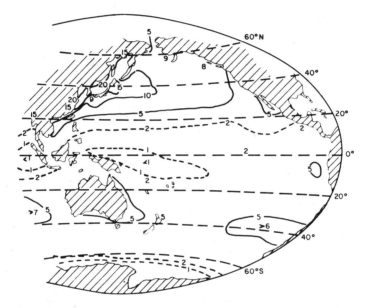

FIG. 4.7. *Amplitude of annual variations of sea surface temperature, Pacific Ocean, in kelvins.*

Figure 4.7 shows the annual variation of surface temperature over the Pacific as an example. Annual variations at the surface rise from 1 to 2K at the equator to between 5 and 10K at 40° latitude in the open ocean and then decrease toward the polar regions (due to the heat required in the melting or freezing processes where sea-ice occurs). Near the coast, larger annual variations (10 to 20K) occur in sheltered areas and particularly in the northwest of the northern oceans for reasons described in Chapter 5. These annual variations in temperature decrease with depth and are rarely perceptible below 100 to 300 m. The maximum temperature at the surface occurs in August/September in the northern hemisphere and the minimum during February/March. Below the surface the times of occurrence of the maxima and minima are delayed by as much as 2 months relative to the times at the surface.

These statements about annual variations of temperature are made on the basis of observations at fixed stations. Defant (1960) has pointed out that in most oceanic regions the surface water is moving and therefore the statements about annual variations at fixed stations refer not to a particular body of water but to a continually changing one. For a region where the ocean currents are zonal, i.e. east–west, there will be little difference between the annual variation at a fixed station and that in a body of water moving with the current. But in a region where the flow is meridional (north–south) this will not be the case because of the difference in annual mean temperature and in annual variation in the north–south direction. For a planktonic organism which drifts with the water it is the variation in the individual water mass which is important. Little attention has been paid in the past to determining the annual change of temperature (or other properties) appropriate to an individual water mass. However, modern satellite tracked, freely-drifting buoys (see Chapter 6) measure surface temperature and thus will provide a means for observing temperature changes while moving with the water mass in which they float.

Diurnal temperature variations at the surface are small in the open ocean (rarely more than 0.3K) but may be larger (2 to 3K) in sheltered and shallow water near the coast. The smallness of the change is partly because there is usually some mixing of the warmed water with cooler subsurface water. A more important factor is that most of the daily heat input from solar radiation is used to evaporate water, leaving only a part available for raising water temperature. Where there is ice in the water, most of the incoming heat will be used to melt it. Diurnal variations extend to depths of only a few metres.

4.34 Deep water; potential temperature

In the deep water below the thermocline zone the temperature generally decreases as depth increases to about 4000 m, the average depth of the world ocean. In the deep trenches, however, the *in situ* temperature often increases slowly with depth beyond 3000 to 4000 m due to the effect of increasing pressure (see Chapter 3). When considering oceanic situations where considerable changes of depth of water masses occur, it is best to plot potential temperature in order to eliminate the effect of change of depth (i.e. pressure) which appears in the *in situ* temperature. An excellent example of this is shown in Table 4.1 and in Fig. 4.8 taken from data of the Dutch *Snellius* Expedition (van Riel, 1934). Figure 4.8(a) shows a sample vertical profile of *in situ* temperature T while Fig. 4.8(b) shows the profile of potential temperature θ. It is seen that while T reaches a minimum at 3500 m and thereafter increases, θ decreases to the bottom. (The salinity changes by only 0.02 ‰ between 3500 m

TABLE 4.1

Comparison of in situ *and potential temperatures, etc., in the Mindanao Trench near the Philippine Islands*

Depth (m)	Salinity (‰)	Temperature		Density	
		in situ T°C	Potential θ°C	σ_t	σ_θ
1455	34.58	3.20	3.09	27.55	27.56
2470	34.64	1.82	1.65	27.72	27.73
3470	34.67	1.59	1.31	27.76	27.78
4450	34.67	1.65	1.25	27.76	27.78
6450	34.67	1.93	1.25	27.74	27.79
8450	34.69	2.23	1.22	27.72	27.79
10035	34.67	2.48	1.16	27.69	27.79

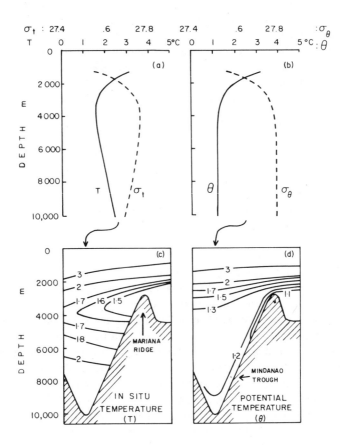

FIG. 4.8.   *In situ and potential temperature and density distributions in the Mindanao Trench: (a, b) vertical profiles, (c, d) vertical sections.*

and the bottom.) The effect of the correction from *in situ* to potential temperature is more dramatically shown in Fig. 4.8(c) and (d) from the Expedition Report. The plot of the *in situ* temperature section (Fig. 4.8(c)) would suggest a flow of cool water over the sill, continuing to descend slowly across the Trench but staying near mid-depth and leaving warmer bottom water undisturbed. The plot of potential temperature (Fig. 4.8(d)) shows a very different pattern with the water which passes over the sill in reality flowing down the slope to the bottom of the Trench.

The profile of σ_t in Fig. 4.8(a) shows a maximum at about 4000 m and then a decrease to the bottom, giving the appearance of instability (more dense water over less dense). However, the potential density (Fig. 4.8(b)) increases to 6450 m and then remains constant showing that when the adiabatic compression with increase of depth is taken into account the water is not unstable but is in neutral equilibrium below 6450 m.

It should be noted that sometimes, in deep water, even a plot of σ_θ may show a decrease with increase of depth, i.e. apparent instability (e.g. Fig. 7.9) because the procedures for calculating σ_θ (and σ_t) neglect the effect of pressure on compressibility, i.e. the values calculated are $\sigma_{S, \theta, 0}$ or $\sigma_{S, t, 0}$. In very deep water it is better to calculate sigma values for a greater pressure. For example, if calculations are made corresponding to a depth of 4000 m, sigma values are shown as σ_4, potential temperature being used (e.g. Fig. 7.10 which no longer shows the apparent instability in Fig. 7.9 below 3500 m). This matter is discussed in more detail by Lynn and Reid (1968) and Reid and Lynn (1971).

4.4 Salinity Distribution

4.41 Surface salinity

The salinity of the surface waters is basically zonal in distribution (Fig. 4.9) although not as clearly so as the temperature. The average surface salinity distribution (Fig. 4.1) is different from that for temperature in that it has a minimum just north of the equator and maximum values in the sub-tropics at about 25°N and S of the equator. The minimum and maxima are evident in the individual oceans in Fig. 4.9. Values decrease toward high latitudes. Observations make it clear that surface salinity is determined by the opposing effects of evaporation increasing it and precipitation decreasing it as is shown in Fig. 4.10. The salinity maxima of Figs. 4.1 and 4.10 are in the trade wind regions where the annual evaporation (E) exceeds precipitation (P), so that (E–P) is positive, while the temperature maximum is near the equator because the balance of energy into the sea has a single maximum there (e.g. see Fig. 5.7(f)).

It will be noted from Fig. 4.1 that the density of the surface water has a single minimum at low latitudes, corresponding to the single temperature maximum.

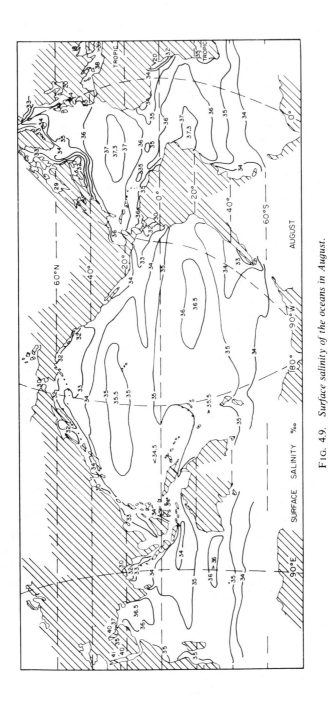

Fig. 4.9.   *Surface salinity of the oceans in August.*

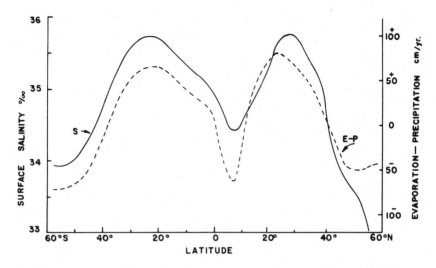

FIG. 4.10.   *Surface salinity (S, average for all oceans) and difference between evaporation and precipitation (E − P) versus latitude.*

The salinity does exert some influence on density but not sufficient for the tropical maxima to appear on the σ_t curve as density maxima.

The range of surface salinity values in the open ocean is from 33 to $37°/_{oo}$. Lower values occur locally near coasts where large rivers empty and in the polar regions where the ice melts. Higher values occur in regions of high evaporation such as the eastern Mediterranean $(39°°/_{oo})$ and the Red Sea $(41°/_{oo})$. On the average the North Atlantic is the most saline ocean at the surface $(35.5°/_{oo})$, the South Atlantic and South Pacific less so (about $35.2°/_{oo}$) and the North Pacific the least saline $(34.2°/_{oo})$.

4.42 Upper-layer salinity

The salinity distribution in the vertical direction cannot be summarized quite as simply as the temperature distribution. In the upper water the reason for this is that density, which is the factor responsible for determining the stable position of a water body in the vertical direction, is determined chiefly by temperature in the open ocean (except in the polar sea). Therefore, water of higher temperature (lower density) is generally found in the upper layers and water of lower temperature (higher density) in the deeper layers. The variations in salinity which occur in open ocean waters are usually not sufficient in their effect upon density to override the effect of temperature. Therefore it is possible to have either high or low salinity in the warmer surface and upper layers.

In the vertical distribution in the equatorial, tropical and subtropical regions there is a marked salinity minimum at 600 to 1000 m (Fig. 4.11) with the

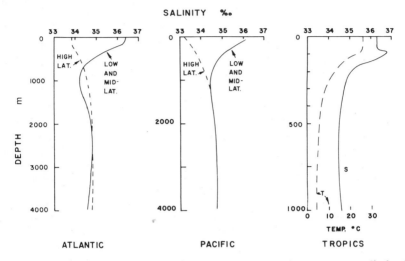

FIG. 4.11. *Typical mean salinity profiles for the open ocean (with temperature profile for the tropics).*

salinity then increasing to 2000 m. In the Atlantic it decreases slightly below this. In the tropics there is often a sharp salinity maximum at 100 to 200 m depth close to the top of the thermocline (see also Figs. 7.26(d) and 7.27(b). This results from water sinking at the tropic salinity maxima (Fig. 4.1) and flowing equatorward. In high latitudes, where the surface value is low, the salinity generally increases with depth to about 2000 m, with no subsurface minimum. In coastal regions, where there is much river runoff, there is generally a zone of rapid increase of salinity, the *halocline*, between the upper, low-salinity water and the deeper more saline water. Here the pycnocline is determined by the salinity distribution rather than by temperature.

4.43 Deep-water salinity

In the deep waters, 4000 m or deeper, the salinity is relatively uniform at 34.6 to $34.9\%_{oo}$ throughout the world ocean. Remembering that the deep-water temperature also has a small range ($-0.9°$ to $2°C$) this means that the deep-water environment is very uniform in character.

4.44 Temporal variations of salinity

Information on temporal variations of salinity is much less than for temperature which is more easily measured. Annual variations of salinity in the open ocean are probably less than 0.5‰. In regions of marked annual variation in precipitation, such as the eastern North Pacific and the Bay of Bengal, and near ice, there are large annual variations. These variations are confined to the surface layers because in such regions the effect of reduced salinity may override the effect of temperature in reducing the sea-water density. This keeps the low salinity water in the surface layer. Diurnal variations of salinity appear to be very small.

4.5 Dissolved Oxygen Distribution

In addition to the solids dissolved in sea-water there are also gases. One which has been widely used as a water characteristic is oxygen, expressed as the number of millilitres of oxygen at NTP dissolved in one litre of sea-water (mL/L). The SI unit of μmol/kg is coming into use but as mL/L are most common in the literature to date, this unit will be used in this text. (Equivalent values for the units are given in the Appendix.)

The range of values found in the sea is from 0 to 8 mL/L, but a large proportion of values fall within the more limited range from 1 to 6 mL/L. The atmosphere is the main source of oxygen dissolved in sea-water and at the surface the water is usually very close to being saturated. Sometimes, in the upper 10 to 20 m, the water is supersaturated with the oxygen which is a by-product of the photosynthesis by marine plants. Below the surface layers the water is usually less than saturated because oxygen is consumed by living organisms and by the oxidation of detritus. Low values of dissolved oxygen in the sea may often be taken to indicate that the water has been away from the surface for a long time, the oxygen having been depleted by the biological and detrital demands.

Figure 4.12 shows typical dissolved oxygen profiles for the Atlantic and the Pacific for three latitude zones. Common features are: (1) the high values close to the surface, (2) the oxygen minimum in the upper 1000 m between the tropics, (3) the relatively high values below 1000 m in the Atlantic (North Atlantic Deep Water), (4) low values in the North Pacific and (5) similar distributions in the southern latitudes in both oceans. Distributions in the Indian Ocean are similar to those in the Pacific (south and tropics). The lower values in the deep water of the Pacific compared with the Atlantic indicate that this water has been away from the surface for a longer period of time and that the deep-water circulation may be slower in the Pacific. In certain regions, such as the Black Sea and the Cariaco Trench (off Venezula in the Caribbean), there is no oxygen but hydrogen sulphide is present instead (from the reduction of sulphate ion by bacteria). This indicates that the water has been stagnant there for a long time.

The conspicuous oxygen minimum in the vertical profiles from between the tropics is apparent in the western Atlantic (Fig. 7.9), in the mid-Pacific

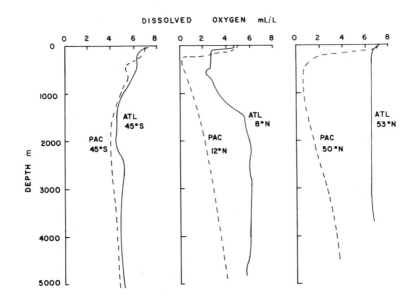

F IG. 4.12.   *Profiles of dissolved oxygen for Pacific Ocean (data from Reid, 1965) and for Atlantic Ocean (data from GEOSECS Atlas, 1976).*

(Fig. 7.34) and is particularly evident in the eastern Pacific (Figs. 7.31 and 7.33) and in the northern Indian Ocean. Suggestions for the cause for this minimum are basically either that this intermediate depth water is in regions of minimal motion so that there is little circulation or mixing to refresh the water and replace the oxygen consumed, or that biological detritus accumulates in this region because of the increase of density with depth (stability) and uses up the oxygen. Neither suggestion is accepted as satisfactory in itself and the oxygen minimum still requires a full explanation. As the production and utilization of oxygen in the sea are essentially biochemical matters they will not be pursued further here but it must be remembered that whenever oxygen is considered as a water property it must be used with caution since it is non-conservative.

4.6 Other Water-Motion Tracers

Other water properties which are used as flow tracers or for identifying water masses are the so-called nutrients, e.g. nitrate, phosphate and silicate ions, dissolved gases other than oxygen, and plankton which are the small organisms which drift with the water (as distinct from the nekton, the free swimming fishes and mammals, which move about of their own volition). These characteristics must be used with caution because, like oxygen, they are non-conservative, i.e. they may be produced or consumed in the water mass, and generally their rates of production and consumption are not well known. Radioactive elements are used as tracers, and the fact they do decay at a known rate is used to estimate the "age" of ocean water as described in Chapter 6.

CHAPTER FIVE

Water, Salt and Heat Budgets of the Oceans.

I N T H E basic sciences much use is made of a number of conservation principles such as conservation of energy, of momentum, etc., and these rather simple principles have very far reaching results and valuable applications. Conservation of heat energy as applied to the oceans will be discussed later in this chapter. First we will discuss two other principles, the second of which is peculiar to oceanography. They are the conservation of volume and the conservation of salt.

5.1 Conservation of Volume

The principle of conservation of volume (or the equation of continuity as it is sometimes called) follows from the fact that the compressibility of water is small. It says that if water is flowing into a closed, full container at a certain rate it must be flowing out somewhere else at the same rate. "Containers" such as bays, fjords, etc., in the oceans are not closed in the sense that they have lids on (except when frozen over), but if one observes that the mean sea level in a bay remains constant (i.e. after averaging out the tides) then there is no flow through the upper surface and the bay is equivalent to a closed container. One might say that this principle is just common sense. This may be true but nonetheless it is science too. It may lead to interesting results. For example, many of the fjords of Norway, western Canada and Chile have large rivers flowing into their inland ends, but on the average the mean sea level in them remains constant. We conclude from the principle of continuity of volume that there must be a simultaneous outflow elsewhere. The only likely place is at the seaward end, and if we measure the currents there we find that in fact there is a net outflow of the surface layer. The direction is correct to balance the inflow from the river but when we check we find that there is a much greater volume flowing out to the sea in this surface layer than in from the river. If conservation of volume is to apply there must be another inflow; the current measurements show that this is from the sea below the outflowing surface layer. The reason for this situation is that the river water, being fresh and therefore less dense than the sea-water of the fjord, stays in the surface layers as it flows toward the sea. However, it picks up sea-water from below en route and the outflowing surface layer includes not only the river water but also the extra salt water picked up. The latter is often in much greater volume than the river water, and the surface outflow to the sea is therefore correspondingly greater than the inflow from the river. In addition, the salt water which has been picked up and flushed out of the fjord must be replaced; this is the cause of the sub-surface inflow from the sea. (This type of circulation is nowadays referred to as an "estuarine" one; it will be discussed in more detail in Chapter 8.)

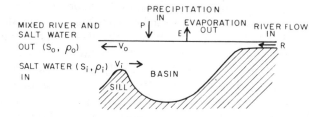

FIG. 5.1.   *Schematic diagram of basin inflows and outflows for Conservation of Volume discussion.*

If we represent these flows schematically as in Fig. 5.1 and add precipitation (P) on to and evaporation (E) from the water surface, the conservation of volume principle may be stated symbolically as:

$$V_i + R + P = V_o + E,$$

or rearranged slightly as

$$V_o - V_i = (R + P) - E = X.$$

Here V stands for *volume transport*, a phrase which is used when we express flow in terms of volume per second (usually m^3/s) rather than as a linear speed (m/s). The second equation simply says that the net volume flow of salt water balances the net volume flow of fresh water (when averaged over a suitable time period). This is an example of a *steady-state* situation in which some or all parts of a system may be moving but at no point is there any *change* of motion (or of property) with time.

5.2 Conservation of Salt

5.21 Principle

The principle of conservation of salt asserts that the total amount of dissolved salts in the ocean is constant. When one first learns that the rivers of the world contribute to the sea a total of about 3×10^{12} kg of dissolved solids per year, the conservation of salt seems to be contradicted. In principle it is, but in practice it is contradicted only to a negligible extent. The total amount of salt dissolved in the ocean waters is about 5×10^{19} kg, and therefore the amount brought in each year by the rivers increases the average ocean salinity by about one part in 17 million per year. But we can only measure the salinity of sea water to an accuracy of about $\pm 0.003 \%_{oo}$, or about 1500 parts in 17 million if we take the mean ocean salinity as $35 \%_{oo}$. In other words, the oceans are increasing in salinity each year by an amount which is only one fifteen-hundredth of our best accuracy of measurement. So for all practical purposes we can assume that the average salinity of the oceans is constant, at least over periods of tens or even hundreds of years. Furthermore, when we apply the principle of conservation of salt to a limited volume where there is no significant input of salt by rivers, the principle applies with even more rigor.

The principle of conservation of salt has been demonstrated above for the world ocean as a whole but in practice it is usually applied to smaller bodies of water. It turns out to be most useful when applied to bodies of water which have only limited connection with the main ocean, e.g. the Mediterranean Sea, a bay or a fjord, as will be demonstrated shortly. Conservation of salt in such water bodies is sometimes taken for granted, but strictly speaking it should be verified before being used. That is, before we use the principle we should determine from observations that the salinity distribution does not change significantly over the period of study. At the same time, there is nothing to prevent us from *assuming* the principle in order to draw some (tentative) deductions. But we must then remember that until conservation of salt has been demonstrated the deductions which depend on it are subject to doubt.

The principle may be expressed symbolically as:

$$V_i \cdot \rho_i \cdot S_i = V_o \cdot \rho_o \cdot S_o$$

where S_i and S_o are the salinities respectively of the inflowing and the outflowing sea-water, and ρ_i and ρ_o the respective densities (Fig. 5.1). Since the two densities will be the same within 3% at the most (the difference between ocean and fresh water) the ρs can in practice be cancelled leaving:

$$V_i \cdot S_i = V_o \cdot S_o.$$

This equation can be combined with the second equation for conservation of volume (Section 5.1) to give Knudsen's relations (Knudsen, 1900):

$$V_i = X \cdot S_o/(S_i - S_o) \text{ and } V_o = X \cdot S_i/(S_i - S_o).$$

One can draw some qualitative conclusions from these relations. In the first case, if both S_o and S_i are large they must be similar (because there is an upper limit to S in the ocean), therefore $(S_i - S_o)$ must be small and both $S_o/(S_i - S_o)$ and $S_i/(S_i - S_o)$ must be large. Therefore V_i and V_o must be large compared with X, the excess of fresh water inflow over evaporation. In the second case, if S_o is much less than S_i, then V_i must be small compared with X while V_o will be only slightly greater than X. For the same value of X for both, the exchange of water with the outside in the first case will be large, while in the second case the exchange will be small. One may therefore expect that the body of water in the first case will be less likely to be stagnant than that in the second case.

5.22 Two examples of applications of the two conservation principles

5.221 The Mediterranean Sea

The Mediterranean Sea is one from which evaporation exceeds precipitation plus river runoff (i.e. for the volume transport equation in Section 5.1, $E > (R + P)$ and X is negative) so that there is a net loss of volume as fresh water which must be made up by inflow of salt water from the Atlantic. As shown in Fig. 5.2(a) the inflow of less saline water through the Strait of Gibraltar is in the upper layer and the outflow is more saline and is deeper because this denser water has sunk from the surface where it was rendered more saline by net evaporation. The two salinity ratios in the equations above for V_i and V_o have values of about 25 which imply that the salt-water flows are both greater by this factor than the fresh-water balance X. Direct measurements of the upper-layer currents give an average value for $V_i = 1.75 \times 10^6$ m^3/s. Then from the equations in Section 5.21, $V_o = 1.68 \times 10^6$ m^3/s and $X = (R + P) - E = -7 \times 10^4$ m^3/s, i.e. evaporation exceeds fresh-water input by 7×10^4 m^3/s. The value above for V_i implies an inflow of 5.5×10^4 km^3/year and at this rate it would take about 70 years to fill the Mediterranean (3.8×10^6 km^3 volume). This may be taken very roughly as a measure of the *residence time*, i.e. the time required for replacement of all the Mediterranean water (sometimes called *flushing time*). The saline outflow at depth, represented by V_o is an important source of salinity for the mid-depth waters of the North Atlantic.

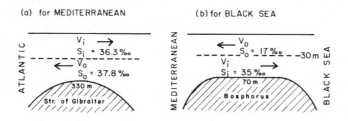

FIG. 5.2. *Schematic diagram of inflow and outflow characteristics for: (a) Mediterranean Sea, (b) Black Sea.*

5.222 The Black Sea

For the Black Sea (Fig. 5.2(b)) the salinity ratios above are 1 and 2 respectively, indicating that here the salt-water flows V_i and V_o are of the same order as the fresh-water balance X. Measured values are approximately $V_i = 6 \times 10^3$ m^3/s   and $V_o = 13 \times 10^3$ m^3/s,   giving $X = (R + P) - E = 6.5 \times 10^3$ m^3/s, i.e. there is a net inflow of fresh water to the Sea. In this case, the value of V_i implies an inflow of saline water of 0.02×10^4 km^3/year. Compared with the Black Sea volume of 0.6×10^6 km^3, this suggests a residence time of about 3000 years.

These residence or flushing-time calculations are very rough but the contrast between 70 years for the Mediterranean and 3000 years for the Black Sea is notable. Oceanographic studies support the contrast as the bulk of the Mediterranean water has an oxygen content of over 4 mL/L whereas the Black Sea water below 200 m has no dissolved oxygen but much hydrogen sulphide (over 6 mL/L). The Mediterranean is described as "well flushed" or "well ventilated" whereas the Black Sea is stagnant below 200 m. As will be described in Chapter 7, the physical reason for the ventilation of the Mediterranean is that quantities of deep water are formed by winter cooling at the surface; in the Black Sea, the salinity and density of the upper waters are too low, because of precipitation and river runoff for even severe winter cooling to make it dense enough to sink to replace deep water.

Other examples of results from the use of these conservation principles will be given in the description of the South Atlantic (Section 7.322) and of the Arctic Sea (Section 7.53).

5.3 Conservation of Heat Energy; the Heat Budget

5.31 Heat-budget terms

It has already been stated that the temperature of the ocean waters varies from place to place and from time to time. Such variations are indications of heat transfer by currents, absorption of solar energy, loss by evaporation, etc. The size and character of the variations in temperature depend on the net rate of heat flow into or out of a water body, and calculations of this quantity are referred to as *heat-budget* studies. In what follows, the symbol Q will be used to represent the rate of heat flow measured in joules per second (watts) per square metre, i.e. W/m^2, generally averaged over 24 hours or over 1 year. (The relationship of these SI units to others which have been used is given in the Appendix.) A subscript will be used to distinguish the different component of the heat budget. These components are:

Q_s = rate of inflow of solar energy through the sea surface,

Q_b = net rate of heat loss by the sea as long-wave radiation to the atmosphere and space,

Q_h = rate of heat loss/gain through the sea surface by conduction,

Q_e = rate of heat loss/gain by evaporation/condensation,

Q_v = rate of heat loss/gain by a water body due to currents which are usually in the horizontal direction so that Q_v is measured through a vertical area of one square metre. The transfer of properties by current flow is called *advection* to distinguish it from the transfer due to diffusion; Q_v is called the *advective term*.

Other sources of heat inflow, such as that from the earth's interior, change of kinetic energy of waves into heat in the surf, heat from chemical or nuclear reactions, etc., are small and can be neglected. The heat budget for a particular body of water can then be stated by the equation:

$$+ Q_s + Q_b + Q_h + Q_e + Q_v = Q_T$$

where Q_T is the resultant rate of gain or loss of heat of the body of water as in Fig. 5.3 which also gives some average values for the Q terms.

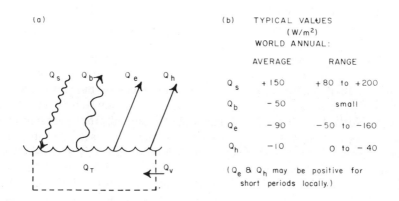

(b)	TYPICAL VALUES (W/m²) WORLD ANNUAL:	
	AVERAGE	RANGE
Q_s	+ 150	+80 to +200
Q_b	− 50	small
Q_e	− 90	−50 to −160
Q_h	−10	0 to − 40

(Q_e & Q_h may be positive for short periods locally.)

FIG. 5.3. *(a) Schematic diagram showing the heat-budget components, (b) typical global annual average and range values for surface terms.*

The above is a symbolic equation. When it is used for heat-budget calculations it is necessary to enter numerical values with a positive sign if they represent gain of heat by the water or with a negative sign if they represent loss from the sea. In practice, Q_s values are always positive, Q_b values are always negative, Q_h and Q_e values are generally negative but may be positive in limited areas at times. Q_v may be positive (inflow of warm water or outflow of cold) or may be negative (inflow of cold water or outflow of warm). Furthermore, as water is substantially incompressible, there will generally be balanced inflow and outflow volumes from a particular sea region (continuity of volume) and the advection of heat by both must be taken into account. (Note also that Q_s, Q_b, Q_h and Q_e are in W/m² and must be multiplied by the sea surface area (m²) of the body of water being considered and Q_v must be multiplied by the vertical area through which advection occurs in order to obtain the total heat flow rate Q_T in watts (W) into (+) or out of (−) the water to cause its temperature to change.)

If the temperature of a body of water is not changing, this does not mean that there is no heat exchange. It simply means that the algebraic sum of the terms on the left of the heat-budget equation is zero—net inflow equals net outflow, an example of a steady-state condition.

If we apply the heat-budget equation to the world ocean as a whole, Q_v will be zero because then all the advection is internal and must add up to zero. Also if we average over a whole year or number of years the seasonal changes average out and Q_T becomes zero. The equation then simplifies to:

$$Q_s + Q_b + Q_h + Q_e = Q_{sfc} = 0.$$

Note that the typical values in Fig. 5.3 for these four terms are only intended as an indication of the general range of annual average values (see Budyko, 1974) and must not be used for specific calculations. It should also be noted that the amount of data on climate for the oceans is limited and so only rounded-off values are given.

The above are *annual* average values, *monthly* averages show larger ranges. For instance, monthly averages of Q_s vary widely from winter to summer in high latitudes (from 0 to about $300 \text{W}/\text{m}^2$ in polar regions) but vary less at low latitudes. Q_h varies with time and place, having maximum values in the north-western North Atlantic and North Pacific, but is generally the smallest term. It may represent a small gain of heat seasonally in some coastal localities (e.g. where upwelling occurs). Q_e is the second largest term in the heat-balance equation and also has large variations, with values as high as $240 \text{ W}/\text{m}^2$ loss being noted in the north-western North Atlantic in winter. Q_b is the only term which does not vary much with time or place. The reason will be apparent later. The variations in the heat-flow terms in different localities give rise to the temperature characteristics of the regions, and the terms will be discussed individually below.

5.32 Short- and long-wave radiation; elements of radiation theory

Before discussing the radiation terms, Q_s and Q_b, some aspects of electromagnetic radiation theory will be reviewed briefly. First, Stefan's Law states that all bodies radiate energy at a rate proportional to the fourth power of their absolute temperature ($°K = °C + 273°$). This energy is in the form of electromagnetic radiation with a range or spectrum of wavelengths. Second, the concentration of energy is not the same at all wavelengths but has a marked peak at a wavelength λ_m given by Wien's Law; $\lambda_m \cdot T = 2897 \mu m \, °K$, where T is the absolute temperature ($°K$) of the radiating body. For a body at a high temperature the radiant energy is concentrated at short wavelengths and vice versa.

The sun has a surface temperature of some $6000°K$ and radiates energy in all directions at a rate proportional to 6000^4. According to Wien's Law this energy is concentrated round a wavelength of $0.5\,\mu m$ ($1\,\mu m = 10^{-6}$ metre); 50% of this energy is in the visible part of the electromagnetic spectrum (about 0.35 to $0.7\,\mu m$) while 99% is of wavelength shorter than $4\,\mu m$. This energy is referred to as *short-wave* radiation and is the source of the Q_s term in the heat budget. The *long-wave* radiation term Q_b represents the electromagnetic energy which is radiated outward by the earth (land and sea) at a rate depending on the absolute temperature of the earth. Taking an average temperature of $17°C = 290°K$ for the sea, it is radiating energy at a rate proportional to 290^4. This is a very much smaller rate than that for the sun, and as the temperature is lower the wavelength is longer. The wavelength at which the sea radiation reaches its maximum is about $10\,\mu m$ (i.e. in the infra-red); 90% of the sea radiation is in the wavelength range from 3 to $80\,\mu m$ and this is referred to as *long-wave* radiation in contrast to that from the sun which is chiefly less than $4\,\mu m$.

5.33 Short-wave radiation (Q_s)

5.331 Incoming solar radiation

A small fraction of the sun's total radiated energy reaches the earth's atmosphere. In Fig. 5.4 this is represented at the top left as 100 parts of in-

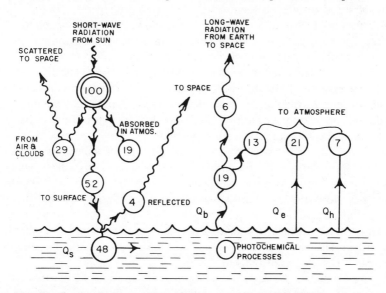

FIG. 5.4.   *Distribution of 100 units of incoming short-wave radiation from the sun to the earth's atmosphere and surface—long-term world averages.*

coming short-wave radiation. Of this about 29 parts are lost to space by scattering from the atmosphere and clouds, 19 parts are absorbed in the atmosphere and clouds, and about 4 parts are reflected from the sea surface. The remaining 48 parts enter the sea. A small part is scattered upward and the remainder constitutes the Q_s term of the heat budget. Of this 48 parts, about 29 parts reach the sea as *direct* radiation from the sun and 19 parts as *indirect* scattered radiation from the atmosphere (*sky* radiation). Note again that this distribution represents a long-term world-area average; instantaneous values vary diurnally, seasonally and with locality and cloud cover.

The rate at which energy reaches the outside of the atmosphere from the sun is called the *solar constant* and is calculated from measurements from the earth's surface to be about $1360\,\mathrm{W/m^2}$ perpendicular to the sun's rays. The 50% which reaches the sea surface then amounts to about $680\,\mathrm{W/m^2}$ when the sun is vertically overhead, or less at other times (due to increased atmospheric absorption).

5.332 Effects of atmospheric absorption, solar elevation, clouds, etc.

The rate at which short-wave solar energy enters the sea, Q_s, depends upon a number of factors discussed in the following paragraphs.

The first factor is the length of the day, i.e. the time during which the sun is above the horizon, which varies with the season and the geographic latitude. In the following discussion this factor has been taken into account wherever possible for figures quoted for the heat-budget terms, and values given are to be understood to refer to the average over a 24-hour period.

The second factor affecting Q_s is absorption in the atmosphere. This depends on the absorption coefficient for short-wave radiation and on the elevation of the sun. The absorption is the combined effect of that due to gas molecules, to dust, to water vapour, etc. When the sun is vertically overhead, i.e. at an elevation of $90°$ above the horizontal, the radiation passes through the atmosphere by the shortest possible path and the absorption is a minimum. When the sun is at an elevation of less than $90°$, the path of the radiation is greater and the absorption therefore greater.

The elevation of the sun has a second effect. If one considers a beam of radiation from the sun of one square metre cross-section this will cover an area of one square metre of calm sea surface when the sun is vertically overhead. At lower elevations, the beam strikes the sea surface obliquely and is distributed over a larger area than one square metre. The energy density, or amount per square metre of sea surface, therefore decreases as the sun moves further from the vertical. The energy density on the sea surface is proportional to the sine of the angle of elevation of the sun.

Figure 5.5 shows the daily inflow of solar radiation at the earth's surface, assuming an average atmospheric transmission of 70% and no clouds, as a

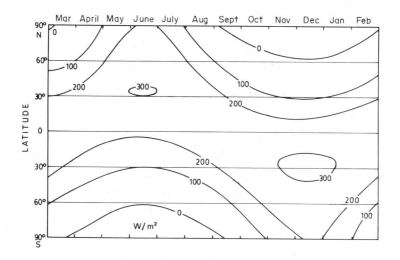

FIG. 5.5. *Daily short-wave radiation Q_s in watts/m^2 received at the sea surface in the absence of cloud.*

function of latitude and time of year. The main features are: (1) the highest values occur at about 30 °N and S latitude in the respective hemisphere summers, (2) there is no short-wave input at high latitudes during the polar winter and (3) the amount of energy input is greater in the southern hemisphere than in the northern. The reason for the last feature is simply that the earth, in its elliptical orbit, is nearer to the sun during the southern summer than during the northern one.

So far in this discussion it has been tacitly assumed that there was no cloud in the sky. The effect of cloud is to reduce the average amount of energy reaching the sea surface below it because of the absorption and scattering by the cloud. The effect of the cloud may be taken into account by multiplying the mean energy which would arrive in the absence of cloud by a factor $(1 - 0.09C)$ where C is the proportion in eighths (oktas) of sky covered by cloud as seen in plan view. For example, if the sky were completely overcast, the value of $C = 8$ and the factor would be $(1 - 0.09 \times 8) = 0.28$, the effect of the cloud then being to reduce the value of Q_s to 0.28 of the clear sky value. For a sky half-covered with cloud, $C = 4$ and the factor would be 0.64

In addition to direct sunlight, the sea also receives a significant amount of energy from the sky, i.e. sunlight scattered by the atmosphere, clouds, etc. The skylight component increases in importance at high latitudes. For instance, at Stockholm (59 °N) for a clear sky in July about 80 % of Q_s will be direct sunlight and only 20 % skylight. In December, only 13 % will be direct sunlight and 87 % skylight. It must be remembered, however, that the total amount of energy reaching the ground will be less in December than in July, and the 87 %

of skylight in December will represent a smaller energy flow than the 20% in July.

The next factor affecting the incoming short-wave radiation is reflection at the sea surface. This depends on the elevation of the sun and the state of the sea (calm or waves). It is necessary to calculate the effect separately for direct sunlight which strikes the sea surface at a specific angle of incidence and for skylight which comes from all directions. For a flat sea, the amount of reflection depends on the sun's elevation as in Table 5.1. For skylight it is calculated that the average amount reflected is about 8%, leaving 92% to enter the water. These figures are all affected by waves but no very good figures for reflection in their presence are available, and the figures of Table 5.1 are used as the best available. A few percent of the radiation entering the sea may be scattered back to the atmosphere.

TABLE 5.1

Reflection coefficient for sea water

Sun's elevation:	90°	60°	30°	20°	10°	5°
Amount reflected (%):	2	3	6	12	35	40
Amount transmitted into water (%):	98	97	94	88	65	60

The rate Q_s at which short-wave energy enters the sea and is available to raise its temperature depends upon all the above factors. Direct measurements of the energy arriving at the sea surface can be made with a pyranometer as described in Chapter 6.

5.333 Spatial and temporal variations

Average values for Q_s allowing for atmospheric absorption, mean cloud amount, scattering, etc., are given by Budyko (1974) with much information on other components of the heat and water budgets also. The annual average value for the short-wave radiation input to the oceans ranges from about 90 W/m^2 at 80°N through a broad maximum of 220 to 240 W/m^2 between 25°N and 20°S, a minimum of 90 W/m^2 at 60°S and rising to 110 W/m^2 at 80°S. Seasonally the ranges of values are greater. In the summer hemisphere, monthly values (averaged over June or December respectively) are from 240 to 300 W/m^2 at 80° latitude, decrease through a minimum of 130 to 150 W/m^2 at 55° latitude, increase to a maximum of 260 to 280 W/m^2 at 30° to 25° latitude, and then decrease to the equator. In the winter hemisphere, at the same time, from the equator values decrease with increasing latitude to about 15 W/m^2 at 60° latitude and to zero poleward of about 70° latitude. The high values in the polar regions are due to the length of the polar summer day and to the low content of water vapour (an absorber) in the atmosphere there.

5.34 Long-wave radiation (Q_b)

5.341 Determining factors

The back radiation term, Q_b, in the heat budget takes account of the net amount of energy lost by the sea as long-wave radiation. The value of this term is actually the difference between the energy radiated outward from the sea surface in proportion to the fourth power of its absolute temperature, and that received by the sea from the atmosphere which also radiates at a rate proportional to the fourth power of its absolute temperature. The outward radiation from the sea is always greater than the inward radiation from the atmosphere and so Q_b always represents a loss of energy from the sea.

The long-wave back radiation Q_b is determined by calculating the rate of loss of long-wave energy outward from the sea from Stefan's Law and subtracting from this the long-wave radiation coming in from the atmosphere. This is measured with a radiometer as described in Chapter 6.

If direct measurements of Q_b are not available it is possible to estimate the heat loss by means of data published by Ångström (1920). He showed that the net rate of loss depends upon the absolute temperature of the sea surface itself and upon the water-vapour content of the atmosphere immediately above it. The temperature of the sea determines the rate of outward flow of energy. The water-vapour content effectively determines the inward flow from the atmosphere because the water vapour in the atmosphere is the main source of its long-wave radiation. Ångström's data were published in the form of a table of values of Q_b as a function of water temperature and of water-vapour pressure. The latter quantity is not measured directly as a rule but instead it is determined from the air temperature and the relative humidity. The latter is easily measured with a psychrometer (wet and dry bulb thermometers). The data were adapted by Lonnquist (Laevastu, 1963) to enable Q_b to be read off a graph (Fig. 5.6) which is entered with the sea surface temperature and the

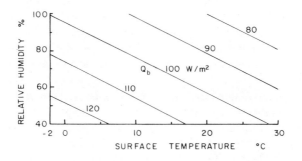

FIG. 5.6.   *Back radiation Q_b from a water surface as a function of surface temperature and relative humidity above the surface in absence of cloud.*

relative humidity. In preparing this graph it was assumed that the air temperature just above the sea surface will be substantially the same as the sea surface temperature, as is usually the case. The values of Q_b for oceanic conditions (with no clouds) range from 115 to 70 W/m^2 with the higher rates occurring at low temperatures and low humidities and vice versa.

The value of Q_b decreases as the sea-surface temperature increases for the following reason. A rise of sea-surface temperature causes an increase of the outward radiation from the sea but is accompanied by an increase of humidity in the atmosphere immediately above it. The temperature of this lower atmosphere follows that of the sea. However, the amount of water vapour increases exponentially, i.e. more rapidly than the temperature, with the result that the atmosphere's radiation into the sea increases more rapidly than the sea's outward radiation. The net result is a decrease of Q_b (i.e. a reduced loss from the sea) as the sea temperature increases. If only the atmospheric humidity increases, the inward radiation from the atmosphere to the sea increases and therefore Q_b decreases.

5.342 Temporal and spatial variations

The back radiation term does not change much either daily or seasonally or with location because it depends on the absolute temperature, not the Celsius temperature, and because the relative humidity does not change much over the sea. For instance, a seasonal change of sea temperature from 10° to 20°C would give rise to a change of outward radiation proportional to $293^4/283^4$ or about 1.15, i.e. only a 15% increase. At the same time the atmospheric radiation inward would increase and reduce the net rate of loss below this figure. The small seasonal and geographic changes of Q_b are in contrast to the large changes of Q_s.

5.343 Effect of clouds

The above values for Q_b are all for the clear sky condition. In the presence of cloud the incoming radiation is increased so that the net rate of loss, Q_b, decreases. The effect of cloud may be allowed for by multiplying the clear sky values by the factor $(1 - 0.1C)$ where C is the amount in oktas of sky covered by substantial cloud (i.e. thin cirrus cloud is less effective than thicker cloud like cumulus). From this it is seen that with the sky completely covered with cloud ($C = 8$) the value of the factor is 0.2, i.e. the loss of energy as long-wave radiation is sharply reduced by cloud cover. This effect of cloud is well known on land where the frost which results from radiation cooling (i.e. the Q_b term for land) is more frequent on clear nights than on cloudy ones. The reason for the big difference between clear and cloudy conditions is that the atmosphere, particularly its water-vapour content, is relatively transparent to radiation in the range from about 8 to 13 μm which includes the peak of the radiation spectrum for a body of the temperature of the sea. In clear weather, energy between 8 and 13 μm wavelength radiated by the sea (and the land) passes through the wavelength "window" in the atmosphere and out into space where it is lost from the earth system.

It should be noted here that for energy in this long-wave part of the spectrum, water has a very high absorption coefficient. The incoming long-wave radiation from the atmosphere is all absorbed, not in the top metres of the sea but in the top millimetres. Similarly the outward radiation is determined by the temperature of the literal surface or skin of the sea. In practice the "sea-surface" temperature as measured is that of a bucket of water dipped from the upper half-metre or so. If the sea surface is disturbed by wind and waves the *bucket temperature* is assumed to represent the skin temperature but very little work has been done in this skin layer. It may seem rather trivial at first to go out to study the top millimetre of the sea, but apart from the heat-budget aspect it is probable that other processes such as the early stages of wind generated waves are determined by stresses in this surface skin. Also, as the upper layer of the ocean is often well mixed vertically this skin temperature, which may be sensed remotely from aircraft or satellites, may be representative of that entire layer.

5.344 Effect of ice and snow cover

When the sea surface becomes covered with a layer of ice, and especially if snow covers the ice, there is a marked change in the heat-radiation budget. For a water surface, the average proportion of short-wave radiation (sun + sky = Q_s) reflected is relatively small (10 to 15 %) and the proportion absorbed is therefore large. For ice or snow, the proportion of short-wave radiation reflected is much larger (50 to 80 %) leaving a smaller proportion to be absorbed. However, the size of the Q_b loss term is much the same for ice as for water, and the result is a smaller net gain $(Q_s - Q_b)$ by ice and snow surfaces than by water. In consequence, once ice forms it tends to be maintained. It has been estimated that the balance in the Arctic Sea is relatively fine and that if the sea-ice were once melted the increased net heat gain $(Q_s - Q_b)$ might maintain the Arctic Sea free of ice (Donn and Shaw, 1966). This, however, would increase the amount of evaporation and there might be marked increases in precipitation on the high north latitude lands which at present receive a relatively small precipitation (mostly as snow in the winter). It should be emphasized that this latter idea about possible changes in the Arctic is very speculative as it is difficult to be sure that the factors which might be changed have been correctly assessed, or indeed that all the factors that would affect the situation have been included. However, it is often by making predictions and then observing their success or otherwise that we test our understanding of natural processes.

5.35 Heat conduction (Q_h)

5.351 Eddy conduction

When we start to consider the conduction term Q_h a new aspect of oceanography comes into the picture. The reason that heat may be gained or lost from the sea surface is that there is often a temperature gradient in the air above the sea. If the temperature of the air decreases upward from the sea surface, heat will be conducted away from the sea and Q_h will be a loss term. If the temperature decreases downward toward the sea surface, heat will be conducted down into the sea and Q_h will be a gain term. In principle the rate of loss or gain of heat is equal to the temperature gradient in the air multiplied by a coefficient of heat conductivity K and the specific heat of air at constant pressure C_p, i.e. $Q_h = -C_p \cdot K \cdot dT/dz$. In small volumes, where the air as a whole is stationary, the process of heat conduction is due only to the random thermal motions of the air molecules (provided that convection is avoided). The coefficient K is then referred to as that for "molecular" conductivity of heat. This quantity is a constant for a particular gas at a particular temperature. However, in nature, and in particular over the sea surface, the air is usually in motion (wind) and the motion is turbulent. A consequence of this is that the air eddies, which consist of bulk movements of the air, carry air properties with them. These properties then tend to move down the property gradient, e.g. heat from higher to lower temperature, but at a much greater rate than is the case when movements of molecules only are concerned. It is possible to define empirically a quantity called the *eddy conductivity* for heat (A_h) where there is turbulent motion, and to describe the rate of heat transfer by the product of this coefficient and the temperature gradient, i.e. $Q_h = -C_p \cdot A_h \cdot dT/dz$. It is the introduction of this concept of *eddy transfer* which is the new idea referred to at the start of this paragraph. It has been introduced here to apply to the conduction of heat through the air above the sea. It applies equally to the transfer or diffusion of water vapour through the air as will be described in the section on Q_e. It also applies to the conduction of heat and the diffusion of salt through the sea where the motion is usually turbulent. Under turbulent conditions the eddy transfer rate for a property is generally so much larger than the molecular rate that the latter may be neglected.

The reason why the introduction of eddy conduction introduces some difficulty is that the eddy conductivity A_h is not a constant quantity, even at a constant temperature, as is the molecular conductivity K. The eddy conductivity depends on the character of the turbulence in the air. This, in turn, depends on various factors such as wind speed and the size of ripples or waves on the sea surface. We do not yet know enough about turbulence in nature to be able to say with any certainty what the value of the eddy coefficient will be in every situation. All that we can really say is that if we happen to have measured the eddy conductivity on a previous occasion when the wind conditions, etc., were much the same, then the eddy conductivity will probably be much the same. Since the determining factor is the character of the turbulence which is impossible to judge by eye and not easy to measure, even with sophisticated instruments, this leaves us in an unhappy situation. It would not be so bad if the eddy conductivity only varied by a small amount with wind, etc., conditions; unfortunately it may vary over a range of ten- or a hundredfold. This whole problem of eddy transfer of properties in fluids is a part of the more general problem of turbulence which is one of the more pressing problems of physics requiring solution at the present time. Research on various aspects is under way and we are gradually acquiring some understanding of this aspect of fluid mechanics. (The eddy transfer process is discussed in more detail in Pond and Pickard, *Introductory Dynamic Oceanography*, in terms of the transfer of momentum, i.e. fluid friction and eddy viscosity.)

However, if we are to study the heat budget we need some way for estimating the heat-conduction term Q_h. It will be shown that it can be related to the evaporation term Q_e by the relation $Q_h = R \cdot Q_e$, where R is Bowen's Ratio, as described at the end of Section 5.363.

5.352 Convection

Before going on to this, some points about Q_h will be mentioned. In situations where the sea is warmer than the air above it, there will be a loss of heat from the sea because of the direction of the temperature gradient. However, the phenomenon of convection will also play a part in assisting the transfer of heat away from the sea surface. Convection occurs because the air near to the warm sea gets heated, expands, and rises carrying heat away rapidly. In the opposite case where the sea is cooler than the air, the latter is cooled where it is in contact with the sea, becomes denser, and therefore tends to stay where it is and convection does not occur. The consequence is that for the same temperature difference between sea and air, the rate of loss of heat when the sea is the warmer is greater than the rate of gain when the sea is the cooler. In the tropics the sea is generally warmer than the air, on the average by about 0.8 C°, and the result is that Q_h is a loss term. In middle and high latitudes the temperature difference is more variable but on the whole the sea is warmer than the air and consequently here also Q_h is generally a loss term.

5.36 Evaporation (Q_e)

Finally, the evaporation term Q_e is an important term but is not easy to determine. Basically there are three methods used to determine it. Two depend on measurements of the rate of evaporation of water while the third is a difference method.

5.361 Pan method

The reason why evaporation enters into the heat budget is that for evaporation to occur it is necessary either to supply heat from an outside source or for heat to be taken from the remaining liquid. The second is the more usual case for the sea. (It is the reason why one often feels cold when one stands with one's wet body exposed to the wind after swimming.) Therefore evaporation, besides implying loss of volume of water, also implies loss of heat. The rate of heat loss is $Q_e = F_e \cdot L_t$, where F_e is the rate of evaporation of water in kilograms per second per square metre of sea surface and L_t is the latent heat of evaporation. For pure water this depends on temperature as $L_t = 2494 - 2.2T$ kJ/kg where the temperature T of the water is measured in degrees Celsius. At 10°C, the latent heat is about 2472 kJ/kg, much greater than the value of 2274 kJ/kg (540 cal/g) at the boiling point. The first two methods for determining Q_e then depend on measuring F_e.

The most obvious way to do this would appear to be to set out a pan of seawater on deck and measure its rate of evaporation by weighing it at intervals, the *pan method*. A better way to determine the water loss is to determine the salinity of the water at intervals and calculate the loss of water from the increase of salinity. There are, however, some difficulties because the surface of the water in the pan is not in the same situation as the sea surface. The pan can be suspended so that the water does not slop over the edge as the ship moves but it is necessary to ensure that no spray gets into it. This necessitates using screens but these at once change the wind flow over the pan and therefore the rate of evaporation. The water in the pan is likely to warm up to a higher temperature than that of the sea itself and this will increase the evaporation rate. A more serious difficulty is that the pan has to be at about deck level which is likely to be several metres above sea level. The water-vapour pressure at this height may be appreciably less than at the sea surface and therefore the rate of evaporation from the pan will be greater than the true rate from the sea. Even on land, in studies of evaporation from reservoirs, it is difficult to get consistent results with pans of different designs. From the few careful studies which have been made at sea it appears that the evaporation rate from a pan at deck level is likely to average about twice the true rate from the sea surface. The average amount of evaporation per year from the ocean is about 120 cm. i.e. the equivalent of the sea surface sinking by that amount due to evaporation. Local values range from an annual minimum of as little as 30 to 40 cm/yr in high latitudes (from free water surfaces, not ice) to maxima of 200 cm/yr in the tropics and decreasing to about 130 cm/yr at the equator. The high values in the tropics are associated with the trade winds while the lower values near the equator result from lower mean wind speeds and seasonally decreased heat input due to greater average cloud cover than in the tropics. The high values for evaporation in the trade wind zones are the reason for the high surface salinity values there (e.g. Figs. 4.1, 4.9).

5.362 Flow method

The pan method being inconvenient and uncertain, it is usual to use another means for estimating the evaporation, the *flow method*. In principle, F_e could be measured by the application of a formula of the type $F_e = -A_e \cdot df/dz$, where A_e is the eddy diffusion coefficient for water vapour through the atmosphere and df/dz is the gradient of water-vapour concentration (humidity) in the air above the sea surface. Unfortunately we are up against the same difficulty for water-vapour diffusion as for heat conduction, because the eddy diffusivity A_e also has a wide range of values depending on the turbulence of the air. In practice a semi-empirical flow formula is frequently used in the form:

$$F_e = 1.4(e_s - e_a) \cdot W \text{ kg per day per square metre of sea surface.}$$

In this formula e_s is the saturated vapour pressure over the sea-water and e_a is the actual vapour pressure in the air at a height of 10 m above sea level, both of these pressures being expressed in kilopascals (101.35 kPa = 760 mm of mercury), while W is the wind speed in metres per second at 10 m height. The saturated vapour pressure over sea-water (e_s) is a little less than that over distilled water (e_s). For a salinity of $35\,°/_{oo}$, $e_s = 0.98e_d$ at the same temperature. The saturated vapour pressure over distilled water may be obtained from tables of physical or meteorological constants. If the water-vapour content in the air is given as relative humidity then the value e_a is equal to the saturated vapour pressure over distilled water at the temperature of the air multiplied by the relative humidity expressed as a fraction, not as a percentage. For example, at an air temperature of 15°C the saturated vapour pressure is 1.71 kPa ($= 12.8$ mm Hg). If the relative humidity is 85% then the actual vapour pressure in the air is $1.71 \times 0.85 = 1.45$ kPa.

This practical formula is basically a simplified version of the theoretical eddy diffusion-flow formula above. In that formula df/dz is the vertical humidity gradient, df being the change in humidity over a vertical distance dz. In the practical formula $(e_s - e_a)$ is the change of humidity over a vertical distance dz of 10 m between the sea surface and the height where e_a is measured. (In the practical formula the factor of 10 does not appear explicitly, having been absorbed in the numerical constant.) The W of the practical formula represents the variation of A_e in an elementary fashion. The values of A_e do not necessarily vary directly as the wind speed but we expect turbulence to increase in some way as wind speed increases and therefore eddy diffusion should increase as wind speed increases. Hence the use of W in the formula does at least give some variation of eddy diffusion in what is certainly the right direction. The actual numerical value of A_e is, of course, not the same as the wind speed; again the factor of proportionality between them is hidden in the numerical constant in the practical formula.

In most regions of the ocean, it turns out that e_s is greater than e_a and therefore as all the other terms in the practical formula are positive, the value of F_e is positive and so is Q_e in these regions. This is entered in the heat-budget equation numerically as a negative quantity as it represents a loss of heat from the sea due to evaporation in such cases. In fact, as long as the sea temperature is more than about 0.3K greater than the air temperature, there will be a loss of heat from the sea due to evaporation. Only in a few regions is the reverse the case, when the air temperature is greater than the sea temperature and the humidity is sufficient to cause condensation of water-vapour from the air into the sea. This results in a loss of heat from the air into the sea. The Grand Banks off Newfoundland, and the coastal seas off northern California are examples of regions where the heat flow Q_e is into the sea (numerically positive). The fogs that occur in these regions are the result of the cooling of the atmosphere.

5.363 Heat-budget method; Bowen's ratio

The third method for estimating Q_e is the *heat-budget method*. If, in the full heat-budget equation of Section 5.31, we consider the situation when $Q_v = 0$ (no advection) and when $Q_T = 0$ (steady state), and we introduce the quantity $R = Q_h/Q_e$ called *Bowen's Ratio* (Bowen, 1926), then $Q_e = (Q_s - Q_b)/(1 + R)$. From this equation, if we have values for Q_s and Q_b and can determine R, we can obtain a value for Q_e. The two radiation terms have already been discussed and it remains to discuss R, the ratio of the heat-conduction term to the evaporation term. Earlier it was explained that the molecular transfer rates for heat and for water vapour are considered negligible by comparison with the eddy transfer rates associated with turbulence. The expressions for $Q_h = -C_p \cdot A_h \cdot dT/dz$ and $Q_e = -L_t \cdot A_e \cdot df/dz$ are similar because the transfer mechanism due to turbulent eddy movements of the air above the sea is of the same physical nature for both. If one goes further and assumes that the numerical values of A_h and A_e are the same, then these two terms will cancel out in Bowen's Ratio, leaving only the ratio of the temperature and the humidity gradients. Each of these gradients may be expressed approximately by the difference of the respective quantity (temperature or humidity) between the sea surface and a level above the sea surface. If the temperature and humidity gradients are both measured over the same range of height dz between the sea surface and some height above sea level (e.g. on the ship's mast or bridge), then the dzs will also cancel in Bowen's Ratio. This then reduces to the simple form: $R = 0.062(T_s - T_a)/(e_s - e_a)$. Here $T_s(^\circ C)$ and e_s (kPa) represent measurements at sea level while T_a and e_a represent the measurements at 10 m height. Both temperature and water-vapour pressure may be measured relatively easily. Hence a value for R is obtained, and this may be used in the earlier equation together with measurements of Q_s and Q_b to obtain a value for Q_e and hence Q_h.

Before going on it should be pointed out that the above simplified practical formula for determining R from the meteorological observations depends on the assumption that A_h and A_e are numerically the same. This is based on the simple argument that the transfers of heat and water vapour are both due to the turbulent motion of the air above the sea. The process of turbulence consists in the transfer of properties from larger to successively smaller eddies and eventually, at the end of the scale, molecular transfer must play a part. While the eddy transfer coefficients for heat, water vapour and other properties tend to have much the same values for high turbulence, they are not numerically equal for low turbulence. The assumption that they are equal when simplifying the Bowen's Ratio formula is therefore an approximation but it seems to be a reasonably good one, judged on the basis of consistency of deductions obtained by its use compared with other evidence.

Average values for R based on many sets of meteorological data are of the order of $+ 0.1$ in equatorial and tropical regions and increase to about 0.45 at 70°N. Remembering that $R = Q_h/Q_e$ this indicates that the heat flow term is usually smaller than the evaporation term. It should be noted that the average values for R are positive because T_s and e_s are usually greater than T_a and e_a respectively. Therefore both of the difference terms in the expression for R are positive. However, it is possible for R to be negative in the regions where the sea temperature T_s is less than the air temperature T_a. A negative value for R indicates the unusual condition of Q_h representing heat flow into the sea.

If a value for R were available, then Q_e could be obtained from the relation $Q_e = (Q_s - Q_b)/(1 + R)$ derived above. This relation, however, depends on the two assumptions that Q_T and Q_v are both zero, and is of limited application. It is not too difficult to check on the first assumption but the second requires detailed current measurements and these are among the most time and effort consuming aspects of experimental physical oceanography if they are to be done well. The heat-budget method for determining Q_e is most useful for large areas for checking heat-budget estimates obtained by other methods. In practice, Q_e is usually obtained from the semi-empirical flow formula, and then with a value for R we can calculate $Q_h = R \cdot Q_e$.

5.37 Geographic distribution of the heat-budget terms

5.371 Pacific Ocean

Annual average distributions of the various Q terms for the Pacific, based on calculations by Wyrtki (1965), are presented in Fig. 5.7(b) to (f). Q_s has a range from $+ 70$ to $+ 220$ W/m^2 (gain) with a strong latitude dependence, related to solar elevations. The low values in the north are due to large cloud amounts there (Fig. 5.7(a)), while high values around 5°S 145°W and 20°N 160°E result from smaller cloud amounts. Q_b (Fig. 5.7(c)) is relatively uniform with a range only from $- 40$ to $- 65$ W/m^2 (loss). The low values in the north are due to large cloud amounts there, and higher values in the north-west are due to conditions to be explained in the next paragraph.

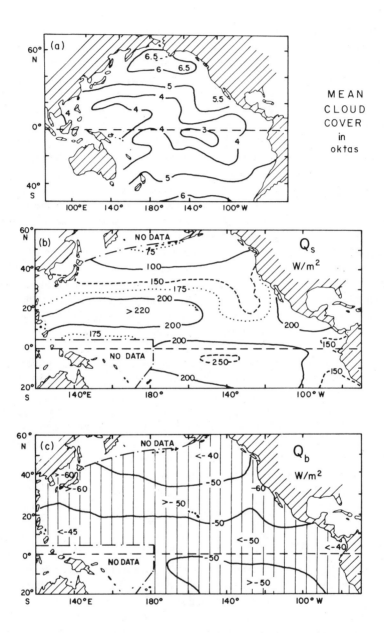

FIG. 5.7.1.   *Pacific Ocean: (a) annual mean cloud cover in oktas; annual mean values in watts/m²*
for (b) Q_s, (c) Q_b. (Note: + values = gain by sea, − value = loss by sea; shading emphasizes loss
areas.)

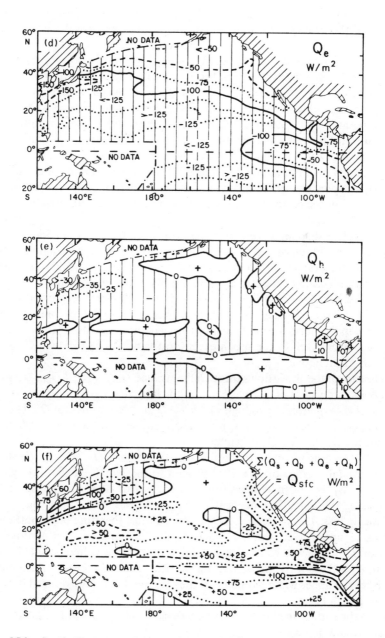

FIG. 5.7.2. *Pacific Ocean: Annual mean values in watts/m^2 for: (d) Q_e, (e) Q_h, (f) Q_{sfc} = (Q_s + Q_b + Q_e + Q_h). (+ values = gain by sea, − values = loss by sea; shading emphasizes loss areas.)*

Q_e, with a range from -25 to over -150 W/m^2 (Fig. 5.7(d)), shows large values in the tropics because $(e_s - e_a)$ is high and there are the steady trade winds to assist evaporation. Even higher values are found in the north-west over the Kuroshio Current off Japan in the winter. The reason for this is that the westward-flowing North Equatorial Current (Fig. 7.24) turns north along the western boundary of the ocean and carries warm water, with relatively high vapour pressure, to higher latitudes where the air temperatures and vapour pressures are lower. This gives rise to a strong humidity gradient resulting in a large upward flow of water vapour from the sea. The maximum values for Q_e occur at the western side of the ocean in winter. The latter fact appears surprising at first, but is simply because the water temperature, and therefore the vapour pressure, of the northward-flowing water does not decrease much in winter, but the temperature and humidity of the air do decrease (cold, dry continental arctic air from the north-west). Consequently, $(e_s - e_a)$ is greater in winter than in summer in this north-western part of the ocean making Q_e larger in winter than in summer. Q_b is largest in winter because of the low humidity of the air. Lowest values of Q_e occur over the upwelling regions (cold water) along the western coasts of the Americas (eastern sides of the ocean).

Q_h (Fig. 5.7(e)), with a range from -10 to $+40$ W/m^2, is generally small over most of the ocean and negative (loss) over large areas. It becomes strongly negative in the north-west off Japan. It reflects the generally smaller values of $(T_s - T_a)$ over much of the ocean except in the north-west in winter, for the reasons given above.

The sum of the above flows through the surface, $Q_{sfc} = (Q_s + Q_b + Q_h + Q_e)$, has high gain values of $+100$ W/m^2 in low latitudes, particularly in the east, south of the equator, due to high values for Q_s and low values for Q_e. The sum has high loss values of -100 W/m^2 in the north-west due to the large negative values for Q_b, Q_e and Q_h in winter.

It is interesting to note that the above distribution of Q_{sfc} (Fig. 5.7(f)) leads to an annual average of 12×10^{14} W gain to the North Pacific through the surface. As the mean temperature of that ocean does not appear to be increasing, conservation of heat energy indicates that there must be a balancing loss. This must be as a Q_v term and must therefore be an outflow of warm, i.e. upper layer, water. This means that there must also be an inflow of subsurface (cool) water. It is calculated that the outflow volume would have to be about 8 Sv (8×10^6 m^3/s), and the vertical speed as the subsurface inflow replaces the upper water would be of the order of 1 cm/day averaged over the whole North Pacific. The location of the outflow is uncertain. Heat may also be leaving the surface in high latitudes where water, in the North Pacific, sinks to intermediate depths.

For the South Pacific, marked differences in the land boundaries and the absence of a strong western boundary current, such as the Kuroshio, may be expected to lead to substantial zonal heat-exchange differences from the North Pacific.

5.372 Atlantic Ocean

The general distribution of the Q terms for the North Atlantic is similar to that described for the North Pacific but there is less information available for the South Atlantic and a direct comparison in detail is not possible.

5.373 World ocean; northern hemisphere

In Fig. 5.8(a) are shown the average annual values for the heat-budget terms in the northern hemisphere. Some of the features revealed by this figure are that the direct sun's radiation (corrected for cloud) dominates to about 50°N but that beyond this the skylight component is equally important, and that while the evaporation-loss term decreases markedly toward the pole, the back radiation term is much the same at all latitudes. The balance between the gain and loss terms (Fig. 5.8(b)) shows a net gain from the equator to 30°N and a net loss beyond this. At first sight there appears to be a much greater loss than gain but this is not really the case. The quantities shown in Fig. 5.8 are the annual average rates of flow *per square metre* of sea surface (W/m^2). To obtain the total flow in or out for any latitude zone one must multiply by the total sea area in that zone. This area is less at high than at low latitudes and with this correction the gain and loss are more nearly, but not exactly, balanced. As there

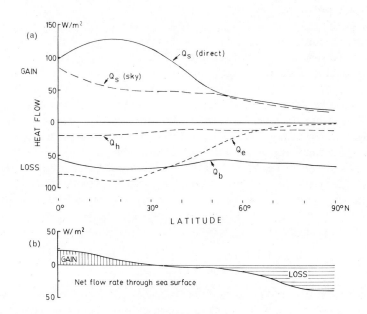

FIG. 5.8. *Values (world ocean average, northern hemisphere) for heat-flow terms through the sea surface as a function of latitude.*

is no indication that the oceans as a whole are getting warmer or cooler we would expect an exact balance. We must conclude that the fault lies in our having insufficiently accurate knowledge of the individual heat-flow terms to permit the budget to be balanced exactly.

For the northern hemisphere as a whole (land and sea), and averaged over the year, Q_s is greater than Q_b between the equator and about 40° N, and there is a net gain of heat by radiation at these lower latitudes. At higher latitudes, Q_s is less than Q_b and therefore there is a net loss of heat by radiation. Since the average temperatures over the earth remain substantially constant we conclude that there must be a net advective flow of heat to the north, from the lower latitudes of net radiation gain to the higher ones of net loss. This heat flow toward the pole is effected both by the ocean and by the atmosphere. These transport warm water or air toward the pole and cooler water or air toward the equator. Recent estimates, based on 9 years of radiation measurements from satellites, indicate that in the northern hemisphere the contribution to the heat transport by the ocean rises to a maximum of 60 % of the total at 20° N, is 25 % at 40° N and 9 % at 60° N (Oort and Vander Haar, 1976). These estimates compute the ocean's contribution to poleward heat transport as the difference between the satellite radiation measurements and observations of atmospheric heat transport from meteorological data. Errors in the atmospheric data may help to explain the large differences between the above residual computation of ocean heat transport and traditional estimates of poleward oceanic transport based on oceanographic observations.

Recent interest in the possible effects of the ocean on the world's climate has prompted a new series of studies into the advective heat fluxes in the ocean. One surprising result has been the suggestion that the oceanic heat flux in the mid-latitude South Atlantic is toward the equator rather than toward the pole. This contradicts conventional thought which requires the South Atlantic to carry heat from the warm equatorial zone to the colder polar latitudes. A summary of the heat and fresh-water fluxes for the world's oceans, based on work by Stommel, is presented in Fig. 5.9. Since the advection (flow) carrying heat in the ocean also carries the particular salinity character of the water, the fluxes of heat and salt are often studied together to provide insight into the accompanying mass or volume flux. The variety in the results of these studies emphasizes both our imperfect understanding of the advective process and the lack of comprehensive data for evaluating the process.

5.374 Large-scale programmes on the heat budget

Apart from the satellite measurements referred to in the previous section, the statements on the heat budget are based on information available before 1964 which is sparse for the northern hemisphere and less than adequate for

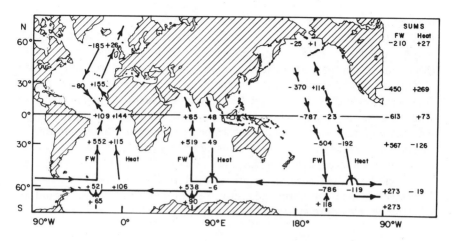

FIG. 5.9.   Fresh-water flux (FW in 10^3 m^3/s) on left, and heat flux (heat in $10^{13} W$) on right, at various latitudes, positive northward.

the southern. The inadequacies in the meteorological data have long been recognized and plans are in hand for acquiring much more comprehensive data. The major developments are expected from the Global Atmospheric Research Programme (GARP) which includes international projects to improve our understanding of the behaviour of the atmosphere. Because so much of the energy which drives the atmospheric circulation comes from solar heat collected by the oceans and returned to the atmosphere as sensible and latent heat, this programme will contain a significant oceanographic measurement component. One of the first major projects under GARP was the GARP Tropical Experiment (GATE) in the eastern tropical Atlantic in 1974 to improve our knowledge of tropical atmospheric interactions, including those with the ocean. A total of thirty-eight ships and twelve aircraft took part in the Experiment over a period of 6 months. A second major effort, called the First GARP Global Experiment (FGGE), took place in early spring and summer of 1979. A fleet of ships from many countries collected a wide variety of oceanographic and atmospheric data in the equatorial regions while many free-floating, satellite-tracked buoys were deployed in the Southern Ocean in an effort to improve significantly the measurement of atmospheric pressure around Antarctica.

Much of the modern interest in the interaction between the ocean and atmosphere comes from the realization that weather cannot be predicted with any skill for periods longer than a few days. The atmosphere is characterized by short term changes and, knowing that most of the energy which drives the atmospheric motions comes from the sea (as latent heat of evaporation), it is hoped that longer-period weather fluctuations, known as climate changes, may somehow be related to the longer thermal memory of the ocean and its interaction with the atmosphere. Efforts by Namias (e.g. Namias, 1972, 1975) have led to the suggestion that the severity of winter patterns in North America may be linked to anomalous (deviations from normal) sea-surface temperature (SST) patterns in the north-east Pacific. In this explanation, abnormally cold SSTs in spring and summer lead to milder winter conditions in western North America and severe cold in the east. Conversely, unusually warm SSTs result in winter storms in the west which then travel north-east leaving eastern North America with mild winter conditions. Imperfections in our understanding of this possible connection keep us from using SST as an operational forecasting tool. Moreover, oceanographers have been unable to explain the mechanisms by which large SST anomalies are created in the north-eastern Pacific. Considerable data collection and study are required if this promising link in the atmosphere/ocean system is to be understood.

CHAPTER SIX

Instruments and Methods

6.1 Introduction

Before describing the techniques and methods of physical oceanography something must be said about the aims and limitations of field observations.

A fundamental goal of many physical oceanographers has been to determine the three-dimensional circulation of the oceans as a function of time. The obvious way would seem to be to go to sea with current meters and to measure it directly. Unfortunately, current meters only give information on the velocity (speed and direction) of the water at the location of the instrument itself, and experience indicates that large variations in current velocity can occur over small distances as well as over small time intervals. Also, in the present state of current meter development it is costly, even with moored instruments, to measure currents even at only a few points at any one time. The limited number of oceanographic ships and of oceanographers available to deploy and recover these current meter moorings also limits the number of spot current measurements that may be made. In consequence, direct measurements of currents have to be restricted to key localities of limited area and for purposes such as testing specific theories. The total of direct current measurements of subsurface currents has provided only a small proportion of our observed knowledge of the ocean circulation.

Failing a sufficiency of direct measurements, the synoptic oceanographer has been forced to use indirect methods. The chief indirect method has been to observe the distributions of water properties, which can be done more expeditiously than observing currents, and to deduce the flow from these distributions. In the majority of cases, this method only reveals the path followed by the water and gives little information on speed. The path is better than nothing but the synoptic oceanographer is always on the look out for any characteristic of the property distributions which will give him an idea of the speed as well as direction. The rate of oxygen consumption has been used in a tentative manner but the built-in clock of radioactive decay offers more promise as it is independent of the physical and biological character of the environment. Carbon-14 (^{14}C), deuterium and tritium, for example, have been used although in the sea their use as clocks is by no means straightforward. They will be discussed in Section 6.28.

The other indirect procedure is to use the geostrophic method. The calculation of currents by this method belongs properly to the field of dynamic oceanography but a brief account will be given later in this chapter because when the principle of the method is understood, the distribution of density, as a water property, can give indications of both the flow patterns and of speed, even if the full dynamic calculation procedure is not carried through. Also the distribution of geostrophic currents, revealed by maps of dynamic topography (to be discussed), has traditionally been an important means for describing the large-scale ocean circulation in descriptive oceanography.

Even when using the indirect methods, the time factor for data collection is significant. The ships available for oceanographic research on the high seas (e.g. Pl. 1) have speeds of only 10 to 15 knots (say 15 to 25 km/h), the distances to be covered are large (thousands of kilometres), and the time taken at each station to sample the water at a sufficient number of depths may be measured in hours. The German research vessel *Meteor* spent 2 years in one study of the South Atlantic alone (e.g. Emery, 1980). For even a small area it may take weeks or months for one ship to complete a survey and if the variations with time are to be studied, years may be required. Some multiple-ship studies have been made of limited areas (e.g. GATE, MODE) and of whole ocean areas (e.g. IIOE) but the organization of such expeditions involves tremendous effort and expense. For a whole ocean it is impracticable to obtain a truly simultaneous picture, although the new satellite observational techniques will help, at least for surface features. The synoptic oceanographer therefore has to make the assumption that, when he analyses them, the data from his cruise or cruises may be considered as simultaneous (sometimes referred to as "synoptic"). It is certainly fortunate that such checks as are available suggest that many of the main features of the open ocean are in a reasonably steady state and therefore the oceanographer's assumption is frequently justified. In fact, it is when he comes into shallow coastal waters that difficulties arise because the variations in properties with position are often greater and the period of change shorter than in the open sea. Also, as a result of MODE and other measurement programmes, oceanographers have realized that the ocean is populated by many small to medium scale (mesoscale) circulation features (*eddies*) analogous to the weather systems in the atmosphere. Proper evaluation of these features requires rapid sampling and the advent of profiling temperature instruments, deployed from aircraft, has helped to provide more nearly "synoptic" pictures of the upper ocean thermal structure over limited areas.

6.2 Instruments

In the following sections some of the basic instruments used in physical oceanography will be described, emphasizing the principles rather than trying to give detailed descriptions.

6.21 Winches, wire, etc.

One of the most essential pieces of equipment on an oceanographic vessel is a *winch* with a drum holding *wire rope* on which instruments are lowered into the sea. For lowering bathythermographs and small instruments, a light-duty winch with some 500 m of 2- to 3-mm diameter steel wire rope and a motor of 1 to 2 kW is used. For water sampling and temperature measurements, a medium-duty winch (Pl. 2) with 2000 to 5000 m of 4-mm diameter wire rope and a 7- to 15-kW motor may be used, while for heavier work, such as dredging, coring, etc., winches with up to 15,000 m of 10- to 20-mm wire and 75 to 150 kW have been used. The wire rope used is multi-strand for flexibility, and made of galvanized or stainless steel (more expensive) to resist corrosion. (Sea-water is one of the most corrosive substances known, given time to act.) The winches must be capable of reeling the wire in or out at speeds up to 100 m/min but must also be controllable in speed so that an instrument can be brought accurately to a position for operation or to where it can be reached for recovery. For instruments which telemeter their information to the surface, steel cable with one or more insulated electrical conductors incorporated is used and the winch must have slip rings to transmit the electrical signals from the wire to the deck instruments while the winch drum is turning.

6.22 Depth measurement

The determination of the depth to which an instrument has been lowered is not always easy. The wire is passed over a *meter wheel* (Pl. 3) which is simply a pulley of known circumference with a counter attached to the pulley to count the number of turns, thus giving a direct indication of the length of wire passed out over it. This length gives the maximum depth which the instrument on the wire has reached. In calm conditions with negligible currents this will be the actual depth. More often the ship is drifting with the wind or surface currents and the wire is then neither straight nor vertical so that the actual depth will be less, sometimes much less, than the length of wire paid out.

The depth of an instrument can be estimated by measuring the hydrostatic pressure at its level, as this is proportional to depth (and to the density distribution in the vertical direction). One pressure-measuring device is a bourdon tube moving the slider of an electrical potentiometer, but this needs an electrical cable to transmit the depth information to the ship. It may be accurate to ± 0.5 to 1 %. Another device is the electrical strain-gauge pressure transducer which uses the change of electrical resistance of metals with mechanical tension. It consists of a resistance wire firmly cemented to a flexible diaphragm, to one side of which the *in situ* hydrostatic pressure is applied. As the diaphragm flexes with change of pressure, the tension in the wire changes and so does its resistance which is measured to provide a value for the pressure and hence for the depth. Accuracies to $\pm 0.1 \%$ or better of full-scale depth range are claimed, with resolution to $\pm 0.01 \%$ or better. The "Vibratron" pressure gauge applies the water pressure to vary the tension in a stretched wire which is caused to vibrate electromagnetically. The frequency of vibration depends on the wire tension and hence on the depth. The vibration frequency is determined to give a measurement of depth to about $\pm 0.25 \%$ accuracy. The use of the protected/unprotected thermometer combination for the estimation of depth is described in the section on temperature measurement.

6.23 Current measurement

There are two basic ways to describe fluid flow, the *Eulerian* method in which the velocity (i.e. speed and direction) is stated at every point in the fluid, and the *Lagrangian* method in which the path followed by each fluid particle is stated as a function of time. In both cases the statements are usually made with respect to axes which are stationary relative to the solid earth. In theoretical studies the Eulerian method is the easier to use, but in describing the circulation of the oceans, as in Chapter 7, the Lagrangian method is used more frequently.

Typical horizontal current speeds in the ocean range from about 200 cm/s (about 200 km/day) in the swift western boundary currents (Gulf Stream, Kuroshio), through 10 to 100 cm/s in the equatorial currents, to a fraction of 1 cm/s in much of the surface layer and in the deep waters. Vertical speeds are estimated to be very much less, of the order of 10^{-5} cm/s or 1 cm/day.

6.231 Lagrangian methods

The simplest Lagrangian current indicator is an object floating in the water with a minimum of surface exposed to the wind. The so-called *drift pole*, a wooden pole a few metres long and weighted to float with only $\frac{1}{2}$ to 1 metre emergent, is often used to determine surface currents close to landmarks. Such a pole is simply allowed to drift with the water, its position being determined at intervals either from the shore or by approaching it in a small boat and fixing its position relative to the shore. Sheets of paper or patches of dye, such as sodium fluorescein, which can be photographed at intervals from a high point of land or from an aircraft are also used.

The drift-pole idea has been extended to the open ocean by using a *freely drifting buoy* (Fig. 6.1) with a radio transmitter so that its position can be determined by radio direction finding from the shore. A more sophisticated and precise drifting-buoy technique, used a great deal in recent years, has been

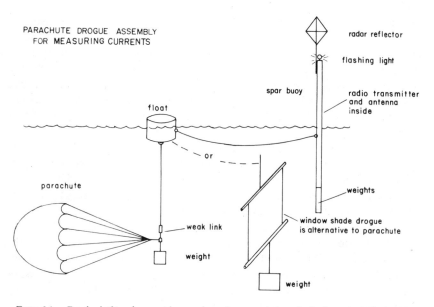

FIG. 6.1. *Freely drifting buoy with parachute drogue; window shade drogue as alternative.*

to fit the buoy with a VHF radio transponder which replies when interrogated from a satellite in orbit. By this means the buoy's position can be determined more accurately than with the lower frequency radio direction-finding technique above. The position of such a *satellite-tracked buoy* is calculated on board the satellite, using the Doppler shift of the buoy's VHF signal, and then transmitted to a ground station for recording. To extend their operating life, such buoys are often fitted with solar panels to keep their batteries charged.

Such buoys are relatively small and, by themselves, may not be perfectly coupled to the water motions. To ensure that the buoys do move with the water, and to minimize the effect of wind, they are frequently fitted with a subsurface drogue to provide additional water drag and more effective coupling with the water motions. As can be seen in Fig. 6.1, this drogue may be in the form of either a parachute or a window shade. The drifting buoys may also be instrumented to measure and transmit surface water properties, atmospheric pressure, etc. These buoys provide both track and speed of flow, i.e. a Lagrangian description, and have revealed many new details of eddies associated with ocean currents.

In the same class as the drift-pole method is the procedure whereby, in fact, most of the information on open ocean surface currents has been obtained. It is a by-product of ship navigation. A ship which is sailing on a given course at a given speed will be deflected from its intended course by surface currents. Therefore a comparison of the actual track (checked by astronomical navigation, landfall, etc.) with that intended will give a measure of the surface current. This comparison may be made every time that a positional fix is obtained from the ship. Maury, about 1853, first suggested examining ship's navigation logs to extract such information on currents, in his case first for the Gulf Stream region off the eastern United States, and the method was subsequently extended worldwide. Most of the maps of surface currents presented in Marine Atlases, and shown in simple form in Chapter 7, are based on the accumulation of such ship-drift data.

To determine the path followed by a pollutant, such as sewage or industrial waste, it is often possible to use the substance itself as a tracer. Samples of water are collected from a grid of positions near the source and in likely directions of flow, and the pollutant concentration determined by chemical analysis. Radioactive materials seem attractive as artificial tracers of water movement, and they were successfully used as a by-product after some of the early Pacific atom-bomb tests. However, in the quantities needed in the sea the cost is often prohibitive, and there is always reluctance expressed by non-oceanographers to the release of radioactive materials in the neighbourhood of communities or commercial fisheries. A very convenient artificial tracer is the red dye rhodamine-B. This can be detected at extremely small concentrations (less than 1 part in 10^{10} of water) by its fluorescence, using relatively simple instruments, and it is also non-toxic at such dilutions. It is only practical to use it in coastal waters, as the quantities required to "tag" open ocean water masses would be impractically large. In all such studies it must be remembered that both advection and diffusion are acting three-dimensionally to spread the tracer and thus results cannot be interpreted solely in terms of advection by currents.

To trace the movements of sub-surface currents, John Swallow of the National Institute of Oceanography in England invented a *neutrally buoyant float* (Swallow, 1955). This "Swallow float" makes use of the fact that the density of the sea increases with depth. The float is adjusted before launching so that it will sink to a selected depth (in terms of density). It then remains at this depth and drifts with the water around it. The float contains equipment to send out sound pulses at intervals and it can be followed by listening to it through hydrophones from the ship. The ship chases the float and at the same time continuously determines its own position. It thereby determines the direction and speed of drift of the float and the water mass in which it is located. A limitation with this instrument is that one ship can follow only a very small number of floats at one time. If several floats are released, e.g. at different levels, it is quite likely that they will be perverse and drift off in different directions and the ship may not be able to keep track of them for long. Recently developed coastal and autonomous moored listening stations make it possible, using the SOFAR technique, to track these neutrally buoyant floats without a ship. As with the satellite-tracked drifting buoys, interpretation is

PLATE 1 (top). *C.S.S. Vector, a 40-m Canadian oceanographic research vessel.*

PLATE 2 (bottom left). *Medium-duty hydrographic winch, hydraulic drive.*

PLATE 3 (bottom right). *Meter wheel with remote indicator.*

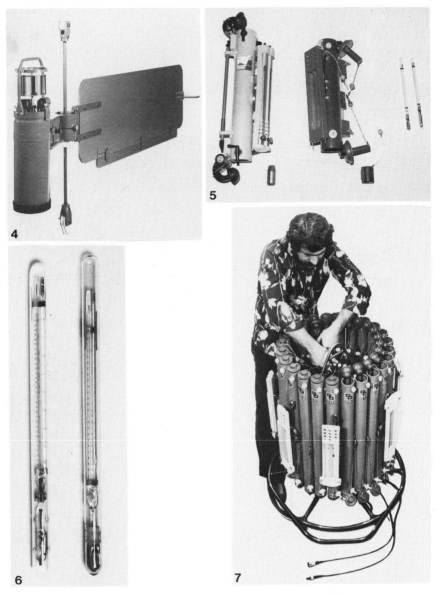

PLATE 4 (top left). *Aanderaa current meter; water property sensors just below current rotor on top of electronics casing.*

PLATE 5 (top right). *Water-sampling bottles (N.I.O. and Niskin), messengers, and protected and unprotected reversing thermometers.*

PLATE 6 (bottom left). *Unprotected and protected thermometers (30 cm long).*

PLATE 7 (bottom right). *Rosette water sampler.*

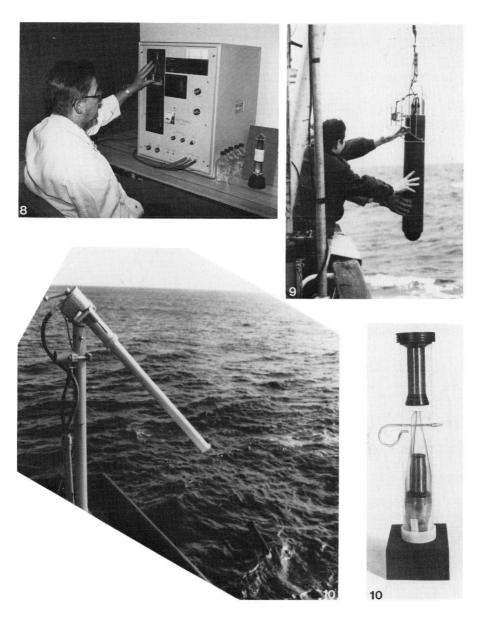

PLATE 8   (top left).   *"Autosal" laboratory salinometer, water-sample bottles and ampoule of Standard Sea-water.*

PLATE 9   (top right). *Sensor head of Guildline CTD.*

PLATE 10   (bottom). (left) *XBT launcher and XBT about to enter water,* (right) *display XBT in transparent casing to show temperature probe location at bottom and dual wire-spools.*

PLATE 11   *Satellite infra-red image off west coast of North America from about 51° to 40° N; white is coolest water, black is warmest.*

not simple since the Swallow floats are not perfectly coupled with the ocean currents. The Swallow float was really the first oceanographic instrument to give us reliable information on the speed and direction of deep currents, and as is usual when one ventures into a new area some of the results obtained have been unexpected (e.g. deep currents in the Atlantic, Chapter 7).

6.232 Eulerian methods; propeller-type meters

The simplest Eulerian current meter is the *Chesapeake Bay Institute drag*. This consists simply of two crossed rectangles of wood (Fig. 6.2(a)) weighted

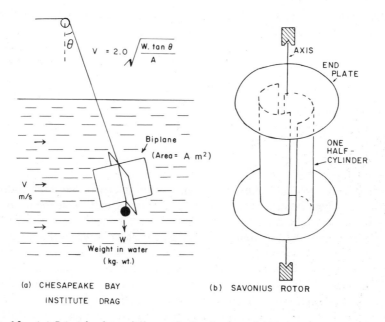

FIG. 6.2.   (a) *Principle of use of Chesapeake Bay Institute drag for current measurement, (b) simple form of Savonius rotor for current meters.*

and suspended by a thin wire. When the drag is immersed, the frictional force of the current pulls the wire to an angle with the vertical. The current speed is related by a simple formula (Fig. 6.2(a)) to the size of the drag, its weight in water, and the angle of the wire from the vertical. This angle is measured to determine the current. This device is simple, cheap to make and quick to use from an anchored ship. It is limited to depths of a few tens of metres because the current drag on the wire increases with length and complicates the interpretation of the wire angle at greater lengths.

Before describing subsurface Eulerian current meters it should be mentioned that one of the fundamental difficulties is determining the *direction component* of the velocity. When the current meter is out of sight below the surface the only frame of reference available to it is the earth's magnetic field. This yields only a small torque to turn a direction indicator and herein lies one of the instrument designer's difficulties. This is particularly so at high latitudes near the magnetic pole where the horizontal component of the earth's magnetic field is small.

Before 1960 the most widely used Eulerian instrument was the *Ekman current meter*. This consisted of a 10-cm-diameter propeller mounted in a frame attached to the end of a wire and lowered to the desired depth. A metal weight (*messenger*) was dropped down the wire to free the propeller to rotate and a second one was dropped after a measured time to stop it, the number of revolutions being recorded by a mechanical counter. The water speed was then proportional to the number of revolutions per minute. The current direction was recorded by the counting mechanism dropping metal balls at intervals along a magnetic compass into a tray with 10° sectors. This instrument had to be lowered and raised for each measurement—a tedious business. An improvement on the Ekman meter was the *Robert's current meter*, the forerunner of most present meters. In this meter, speed (from a propeller) and direction (from a compass) were transmitted electrically to the surface and recorded on shipboard or transmitted by radio from the supporting buoy to a mother ship.

One disadvantage of the propeller-type current meter is that up-and-down motion, as when the ship rolls or the mooring moves, may cause the propeller to turn and cause inaccuracies in the speed measurement. A hollow cylinder with its axis horizontal mounted round the propeller minimizes this effect. An alternative to the propeller is the *Savonius rotor* which is less sensitive to vertical motion. It consists of two half-hollow cylinders mounted on a vertical axis with flat end-plates (Fig. 6.2(b)) and has the advantage of producing a large torque even in small currents. The rotor is made of plastic to be neutrally buoyant to reduce bearing friction so that it is sensitive to currents of as little as 2 cm/s. Even this low threshold value can be a problem in parts of the ocean where currents of this order prevail. The rotor carries several small magnets and as each passes a coil on the frame it induces a momentary electrical current pulse. The number of pulses per second is proportional to the current speed. The current direction is determined electrically with reference to a magnetic compass. The speed and direction information is either transmitted electrically up the supporting cable to the ship and there recorded or, more often, recorded within the instrument on photographic film or magnetic tape. The internally recording arrangement is used for current meters suspended from buoys for long-term measurements; the magnetic-tape system is most convenient for computer analysis of the records.

An example of a modern current meter is the Aanderaa (Pl. 4) which is about

75 cm high and 140 cm long. On magnetic tape it records current speed from 2.5 to 250 cm/s to $\pm 2\%$ or ± 1 cm/s, whichever is the greater, direction to $\pm 5°$ with a magnetic compass, depth to $\pm 1\%$ of range and time to ± 2 s/day for up to a year. It can also record temperature to ± 0.2 C° and conductivity to $\pm 0.1\%$ of the range from 0 to 7 siemens/m which corresponds to salinity from 0 to over $40°/_{oo}$, depending on temperature. Data may also be telemetered up an electrical cable or by acoustic link to the surface so that they may be observed in real time (i.e. as the events occur) to ensure that the meter is working correctly and so that all the data will not be lost if the current meter cannot be recovered.

Many current meters record speed at short intervals but record direction only at longer intervals, which can lead to uncertainties when it is desired to know the (rectangular) components, e.g. to north and east, of the velocity. The *Vector-averaging Current Meter* (VACM) is designed to measure the velocity frequently, resolve it into components and record these separately, to give a more complete record of the velocity. *Vector-measuring Current Meters* (VMCM) use two propellers at right angles to measure the rectangular components directly (relative to a magnetic compass).

6.233 Eulerian methods; non-propeller-type meters

A different method for current measurement is to use the rate of cooling of an electrically heated wire as a measure of the fluid speed past it. This is the *hot-wire anemometer*. A thin wire or metal film about a millimetre long is exposed to the flow and maintained at a constant temperature by automatically adjusting the electric current through it so that the Joule heating is exactly equal to the rate of loss to the fluid. The magnitude of the electric current is then a measure of the fluid speed. This device has the advantage of small size and very rapid response to flow variations which makes it particularly suitable for the measurement of turbulent fluctuations of flow speeds.

Another technique for this purpose is to use a *sonic anemometer* in which the speeds of travel of pulses of high-frequency sound in opposite directions are measured, the difference being a measure of the component of fluid speed along the sound path. There are also *Doppler anemometers* in which the (Doppler) frequency shift of sound or light (laser) reflected from particles in the water is measured to give the water velocity, including short-period fluctuations characteristic of turbulence.

The *electromagnetic method* uses a fundamentally different principle, first suggested by Faraday (1832), that an EMF will be induced in a conductor which moves across a magnetic field. In this case, sea-water is the conductor and when it flows across the lines of force of the earth's magnetic field an EMF, $E = B \cdot L \cdot v$, will be generated where v is the water speed, L the width of the current and B the strength of the earth's magnetic field component in a direction mutually perpendicular to the direction of both v and L. For a horizontal current, B would be the vertical component of the earth's field. Faraday was unsuccessful in applying his idea to measure the flow of the Thames because of problems with copper electrodes; some of the earliest reported measurements by this technique were of tidal currents in the English Channel (Young *et al.*, 1920), and a long series of measurements was made of the Florida Current between Key West and Havana (Wertheim, 1954). The basic equipment required is a recording millivoltmeter and two electrodes to dip in the sea. The electrodes are best placed one on each side of the current and so a further requirement is an insulated connecting wire to the farther electrode. Unused commercial cable circuits have often been used for this purpose. One source of error is the finite, but usually unknown, electrical conductivity of the sea bottom which allows an electrical current to flow due to the induced EMF and so reduces the observed EMF below that to be expected from the formula. This introduces a constant scaling factor which must be determined by making some water-current measurements with another type of meter while the electromagnetic system is in operation.

A requirement for the electromagnetic method is stable electrodes so that electrochemical effects will not complicate the interpretation of the records. Silver wire with silver chloride deposited on it is the most satisfactory electrode for use in the sea.

An adaptation of this technique, by von Arx (1962), to permit underway shipboard measurement was called the *Geomagnetic Electrokinetograph* (*GEK*). Two electrodes were towed behind the ship and the EMF induced in the length of cable between the electrodes was recorded as a measure of the component, perpendicular to the ship's track, of the water velocity. To obtain the total water velocity, the ship was then turned at right-angles to the original track and a second component measured. Combining the two components gave the water velocity relative to the solid earth. The difficulty of reducing and interpreting GEK data led to a rapid decline in its use. A theoretical discussion of both the fixed electrode and the GEK principles was given by Longuet-Higgins, Stern and Stommel (1954).

The small magnitude of the earth's magnetic field together with electrical noise always present in nature makes the geo-electromagnetic method practical only with electrode separations of tens of metres or more. In 1947 Guelke described an electromagnetic current meter in which a local magnetic field was generated by an alterating current in a coil, the field being strong enough to give rise to a measurable EMF with electrodes only 20 cm apart when used to measure currents in Capetown Harbour. The use of an alternating magnetic field permitted amplification of the alternating EMF generated and largely eliminated long-period noise and electrode effects. Recently current meters employing this principle with even smaller electrode spacings have become available commercially. They have the advantage of having no moving parts but do need a significant electrical power supply.

6.234 Mooring current meters and other instruments

Even with the development of improved instruments, the measurement of currents from a ship has several disadvantages. A major one is that any ship movement introduces into the measured currents spurious components which are difficult to determine and to get rid of in analysis. In shallow water it is easy to anchor but wind and tide still cause ship movement. In deep water it is usually difficult to anchor at all, requiring special cables and large winches. Also, the fact that the anchor cable cannot be vertical leaves scope for ship movement which is usually difficult to measure with sufficient accuracy to apply corrections to the current data. The anchor cable may not be vertical because this would mean that the anchor would be pulled out of the bottom by any ship movement, allowing the ship to drift. Anchoring to a weight alone is impractical because an impossibly large weight would be needed. However, a small buoy may be anchored firmly to a weight which is small enough to handle from a ship. In this case, the anchor wire may be taut so that the buoy does not have scope to move significantly. Knauss (1960) developed a technique for continuously manoeuvring the ship so that it was in a known position (by radar) relative to the buoy at all times so that current measurements could be made. Even this technique has the disadvantage that measurements are made at one location only and that the ship and crew are tied up there for the duration of the current measurements at a cost of many thousands of dollars per day. The duration will be for a minimum of 15 days if one wishes to have a long enough time-series to be able to resolve even the major tidal constituents, and often longer series, e.g. a year, are desired.

For these reasons, techniques for the successful mooring and recovery of strings of current meters in the deep ocean, supported on a cable from a buoy beneath the surface to an anchor weight on the bottom, have been developed since the mid-1960s, and most current measurements are now made in this way. There are still problems associated with the movement of the current meter strings in strong currents, and the meters may not always be recovered, but the number of depths at which currents may be measured improves the description of the flow and the length of time-series observations permits resolution of the currents into mean and fluctuating components of various periods.

There are many different schemes for mooring current meters. A variety are shown in Fig. 6.3, most employing a meter similar to the Aanderaa described earlier. An early deployment and recovery technique was to have a double anchor with a float at the surface marking the second anchor. This could be recovered and the system brought in, ending with the current meters and flotation. With this system, subsurface flotation could be used which greatly reduced mooring motion from surface effects. A modification of this system eliminated the surface marker for the second anchor and the ship grappled to find the cable between the two anchors. The absence of a surface marker is

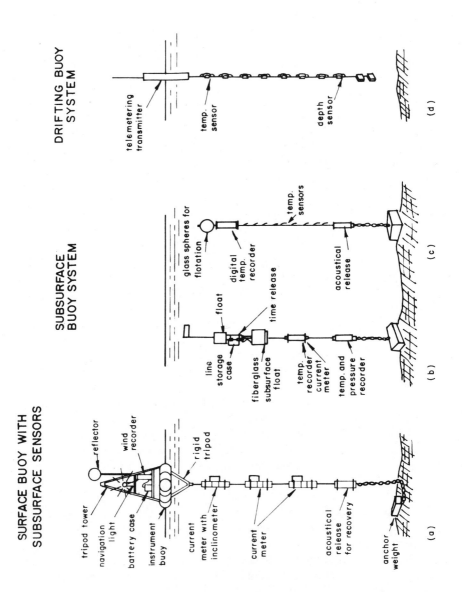

FIG. 6.3. *Schemes for mooring current meters: (a) surface buoy with meteorological instruments and subsurface current meters, (b) subsurface buoy with water property and current measuring instruments, (c) "chain" of temperature sensors with recorder, (d) drifting temperature "chain".*

advantageous when there are ships or icebergs in the area. All of these methods are time-consuming and risky. Release systems were developed that could cut the cable, being actuated either by a timer system or fired by an acoustic signal sent from the ship. Due to weather and scheduling problems, it has become more common to use the command system and the acoustic release has become a usual component of the modern current meter mooring.

As shown in Fig. 6.3, both surface and subsurface flotation systems are used in modern moorings. Frequently surface buoys are needed either for surface meteorological observations, such as wind, pressure, etc., or for data telemetry. Many different types of anchor are used ranging from cement blocks to surplus railroad wheels. Mooring lines themselves have consisted of steel cable, nylon rope and a recent synthetic called Kevlar. Flotation devices also vary widely but one frequently used element is the glass sphere, because it is cheap and will withstand pressure. Often a mooring is deployed in reverse by trailing the flotation and current meter string behind the ship and finally dropping the anchor when the ship is over the desired location.

Recent technological efforts have produced a number of vertical current profiling devices. One, called the *Cyclosonde*, moves up and down a mooring line, measuring currents with a ducted propeller and other water properties. A similar current profiler developed by Düing (Düing and Johnson, 1972) consists of an Aanderaa meter in a streamlined body which is lowered from a ship anchored or drifting. Ship movements are carefully tracked and the current meter record is corrected for ship displacement.

6.235 Geostrophic method and dynamic topography

Most of our information on subsurface currents has been obtained by the use of the *geostrophic method*. Strictly speaking, this is in the domain of the dynamic oceanographer and a full discussion is given in *Introductory Dynamic Oceanography* by Pond and Pickard. We will present the elements here, in a qualitative manner, because it is possible to deduce some of the velocity characteristics from a simple examination of vertical profiles of density (or of temperature in the upper layers) and from horizontal maps of a quantity called *dynamic height* which is related to the density distribution. The geostrophic method is based on a simplified view of ocean dynamics in which friction is neglected and boundaries are assumed to be far away. This results in a balance between the force due to a pressure gradient and an apparent force, the *Coriolis force*, resulting from the rotation of the earth.

To indicate how we apply this situation in oceanography, imagine a particle of water W (Fig. 6.4(a)) in a region where the pressure surfaces p_1, p_2, p_3, etc. (dashed lines) are inclined to the horizontal or level surfaces (full lines). Then the pressure on the right of W will be greater than on the left and there will be a pressure gradient force (PGF) to the left. This would cause it to move to the left

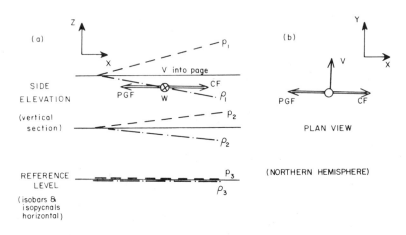

FIG. 6.4. (a) Schematic diagram (vertical section) of isobaric and isopycnal surfaces and corresponding steady-state, no friction, geostrophic flow direction (northern hemisphere) with pressure gradient force (PGF) and Coriolis force (CF) in balance, (b) plan view of the balanced forces and corresponding flow direction (V).

but as soon as it starts to do so, the Coriolis force (CF) will come into play to the right of its velocity V and it will settle down to a steady velocity (into the paper in the side elevation of Fig. 6.4(a); in the y direction in the plan view of Fig. 6.4(b)), with the two forces balancing each other. It can be shown that:

$$V \cdot 2\Omega \sin \phi = -\frac{1}{\rho} \frac{\partial p}{\partial x} \text{ (the geostrophic equation)}$$

where V = speed of flow, Ω = angular speed of rotation of the earth, ϕ = geographic latitude, ρ = water density and $\partial p/\partial x$ = horizontal pressure gradient. The dynamic oceanographer uses this equation, or a variation, for calculating geostrophic speeds, while the meteorologist uses a similar relation for the atmosphere to calculate wind speeds.

Now we do not measure pressure directly as a routine matter in ocean-ographic surveys but we do know that hydrostatic pressure is determined by the density structure of the ocean and we can determine this from measurements of temperature and salinity. In Fig. 6.4 the isopycnal (constant density) surfaces are represented in section by the dash–dot lines. They have the opposite slope to the isobars (constant pressure surfaces). If we plot a vertical section of density, the slope of the isopycnals will tell us whether the current is flowing into the section (as in Fig. 6.4(a)) or out of it (if the isopycnals sloped the other way). In the upper waters, where density is determined chiefly by temperature (mid- and low-latitudes), a temperature section will permit the same estimate of flow direction (as the isotherms will slope the same way as the isopycnals). Also, the steeper the isopycnals/isotherms, the faster the current.

This is one of the uses made by descriptive/synoptic oceanographers of the geostrophic method. A useful rule-of-thumb is that, in the upper layers, in the northern hemisphere the "light water is on the right of the direction of geostrophic flow". (In the southern hemisphere, conversely, "light is on the left".)

If the slopes of the isopycnal surfaces vary with depth, this indicates that the currents vary with depth. We say that, in this case, there is "current shear" or "geostrophic shear". This is characteristic of the *baroclinic* situation in which the slopes of the isopycnals vary because the water density depends on water properties (temperature and salinity) as well as on pressure (depth). The *barotropic* situation is one in which the density depends on depth only, as in an isothermal fresh-water lake. In this case the isopycnal surfaces must be parallel. If the water is stationary they will be horizontal, if the water is moving they will all be at the same angle to the horizontal (determined by the geostrophic relation) and there will be no current shear, i.e. the current will be the same from top to bottom. This is called barotropic flow and the geostrophic equation gives us no information about it. The total current in the sea then may be a combination of baroclinic and barotropic components.

A limitation, in principle at least, to the use of the geostrophic relation is that it only gives us relative currents, i.e. the current at one level relative to that at another. To convert these relative currents into absolute currents, which are what we need, we must determine the absolute current at some level. The most usual way has been to assume that the absolute current is zero at some depth (the *depth of no motion* or *reference level*) and then the geostrophic equation can give us the absolute current at all other levels. In the Pacific, water properties suggest that the water below 1000- to 2000-m depth has very little motion and we can assume a depth of no motion in that depth range. In the Atlantic, on the other hand, as will be shown in Chapter 7, water properties suggest significant motion both in the upper and in the deep and bottom waters, so that a depth of no motion between the two is appropriate.

The reason why the geostrophic method is only relative is that our shipborne measurements are made from the sea surface which may itself be sloping (if there are surface currents as is usually the case) and we have no means at present for determining the slope of the sea surface with the accuracy needed. We can, therefore, only determine the isopycnal slopes relative to one another, and hence the currents calculated will be only relative ones.

Assuming no motion at the reference level implies that at this depth there are no pressure or density gradients, i.e. isobars and isopycnals are level as in the lower part of Fig. 6.4(a). If the total current is a combination of barotropic and baroclinic components, e.g. tidal plus density determined flows, then only the baroclinic component will be given by the geostrophic calculation. Also, it must be noted that although the currents will diminish to zero in contact with the bottom, because of friction, this bottom zone may not be used as a depth of no motion with the geostrophic relation because the latter was derived on the assumption that there was no friction.

An alternative, nowadays, to assuming a depth of *no* motion is to use a depth of *known* motion if there are current meter records or Swallow float observations available for the region, either of which give the absolute velocity at some depth and therefore permit us to convert the relative geostrophic currents into absolute currents. We may still be left with the problem of separating the barotropic and baroclinic components if we wish to understand fully the current regime.

To use the geostrophic equation it is necessary to have a minimum of two oceanographic stations to calculate the isopycnal slopes and hence the currents. In practice, it is usual to select a line of stations and calculate the currents between each pair and then plot a vertical section of the (horizontal) currents. These will only be the components perpendicular to the line of stations of the actual water currents. To derive the total currents, we must use a grid of stations with lines, for example, north–south and east–west and calculate east–west and north–south current components respectively and then combine them to give the total (vector) velocities.

If we wish to study horizontal flow over an area rather than through a vertical section, the usual procedure is to plot a contour map of the elevation of a selected isobaric surface relative to some deeper level, e.g. an assumed depth of no motion. In regions where the water density in the column is low the isobaric surface will be high, and where the water density is high in the column the surface will be low. The result will be a contour map of the isobaric surface, like a topographic map of the land, i.e. a *dynamic topography map*. Flows on this surface will be *along* the contours with the "hills" on the right of the flow in the northern hemisphere (on the left in the southern) and the speed of flow at any point will be proportional to the steepness of the slope at that point, i.e. inversely proportional to the separation of the contours.

The height of the upper isobaric surface will depend on the total density distribution in the column between it and the depth of the reference level below. The quantity used for contouring is called the *geopotential distance* (*dynamic height* in oceanographic jargon) between the surfaces. For two isobaric surfaces p_2 (upper) and p_1 (lower), the geopotential distance is:

$$\Phi_2 - \Phi_1 = \int_1^2 \alpha \cdot dp = -\int_1^2 \alpha_{35, 0, p} - \int_1^2 \delta \cdot dp.$$

The first integral term is called the *standard geopotential distance* ($\Delta\Phi_s$ or $D_{35, 0, p}$) as it depends only on the pressure difference; the second ($\Delta\Phi$ or ΔD) is the *geopotential anomaly* (or *dynamic height anomaly* in older literature) and it contains the effects of the actual temperature and salinity distributions relative to the arbitrary standard values of $0°C$ and $35°/_{oo}$ for the standard distance. In practice we only calculate the anomaly and refer to it as the "dynamic height". (Note that although the quantities are called "distance" and "height", and the latter is quoted in units of "dynamic metres" or "dynamic centimetres", they have physical dimensions of energy per unit mass as they represent work done against gravity. Their correct SI units are $J/kg = m^2/s^2$.) For numerical convenience, oceanographers have used the dynamic metre such that 1 dyn m $= 10.0 J/kg$. To indicate that the dynamic metre (dyn m) is being used, the symbol D is used for geopotential, and the geopotential distance $(D_2 - D_1)$ is then numerically almost equal to the depth difference $(z_2 - z_1)$ in metres, e.g. relative to the sea surface and taking the acceleration due to gravity $g = 9.8 m/s^2$:

	SI units	Mixed units
at a geometrical depth in the sea	$= +100 m$	$= +100 m$
then	$z_2 = -100 m$	$= -100 m$
and the pressure will be about	$p_2 = +1005 kPa$	$= +100.5 dbar$
and the geopotential distance relative to the sea surface	$(\Phi_2 - \Phi_1) = -980 J/kg,$	
	$(D_2 - D_1)$	$= -98 dyn m.$

(See the Appendix for definitions of the pressure units kPa and dbar.)

Two examples of such dynamic topography maps are shown. For the Pacific (Fig. 6.5), Wyrtki (1974) computed the annual average field of dynamic height of the zero pressure surface (sea surface) relative to the 1000 dbar surface (10,000 kPa) ($\Delta D = 0/1000$ dbar) from all available oceanographic stations. For the North Atlantic, Stommel, Niiler and Anati (1978) computed $\Delta D = 100/700$ dbar as shown in Fig. 6.6. (Arrows have been added to the dynamic height contours to assist the reader in interpreting the direction of flow.) Note the similarities, especially the clockwise circulations north of the equator and the strong western boundary currents (contours very closely spaced), and the anticlockwise circulation of the South Pacific bordered on the south by the Antarctic Circumpolar Current. It should be mentioned that while the speeds in the Pacific at the 1000-dbar level are probably quite small in most regions, this is probably not the case at the 700-dbar level chosen for the Atlantic map. Such maps provide considerable detail to a description of part of the current field in the ocean. However, it should be remembered that the observations used were not "synoptic", i.e. not made simultaneously, and any variation from the steady state which occurred during the period covered by the observations would introduce irregularities or "noise" into the maps. In most cases, dynamic topography maps (and plots of water properties) for ocean size areas have to be prepared from data collected by various expeditions, in different parts of the area, and over many years. The fact that well-defined features appear in such maps, or are repeated when there are sufficient data to prepare maps for the same area for several seasons or several individual years, indicates that these must be permanent features of the ocean.

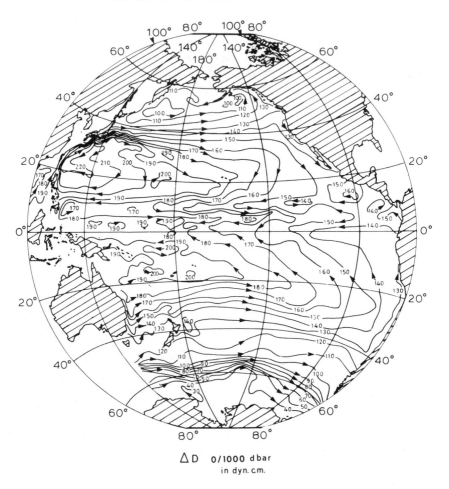

ΔD   0/1000 dbar
in dyn. cm.

FIG. 6.5.   *Mean annual dynamic topography of the Pacific Ocean sea surface relative to 1000 dbar in dyn cm (ΔD = 0/1000 dbar); 36;356 observations.*

6.24 Water properties

6.241 Water-sampling bottles

In order to determine the properties of a sample of sea-water it is necessary first to obtain the sample. For a "surface" sample, a bucket on a rope often suffices to obtain water for temperature and salinity measurement. A plastic bucket is best, as less likely to pollute the sample than a metal one, and the experienced oceanographer will be found using a small one containing a litre or so rather than a full-sized bucket.

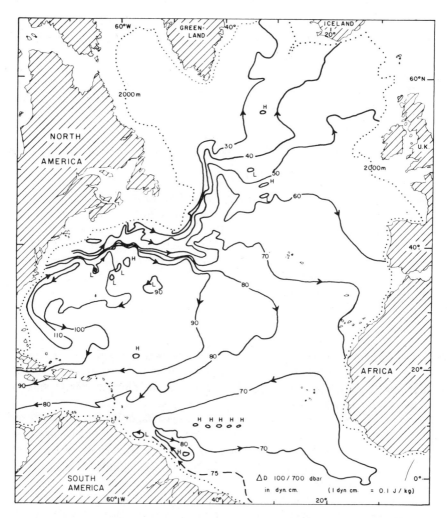

FIG. 6.6. *Dynamic topography of 100-dbar surface relative to 700-dbar surface (ΔD = 100/700 dbar) in dyn cm, Atlantic Ocean.*

For subsurface samples a variety of *water-sampling "bottles"* (e.g. Pl. 5) are available. These are generally metal or plastic tubes with either plug valves at each end (Nansen bottle) or spring-loaded end-caps with rubber washers. The bottle with the ends open is attached to the wire and lowered to the desired depth. There it is closed by the tripping action of a *messenger* (a small metal weight which is slid down the wire. Generally a number of bottles (12 to 24) are attached in series at predetermined intervals along the wire (a *bottle cast*) and closed in succession. (Each in turn releases a messenger to close the next below it.) When the bottles have been brought back on deck the water samples are drawn through a tap, following a routine designed to obtain a pure sample. In some designs, the bottle when tripped is released at its upper end and rotates through 180° about a hinge at its lower end where it is clamped to the wire. This is for the purpose of operating the *reversing thermometers* described later, and leads to the bottles being referred to as "reversing water bottles". In other designs, the bottle remains stationary while a frame carrying the reversing thermometers rotates. A capacity of 1.25 litres is common for these bottles but for special purposes, such as C^{14} analysis, larger bottles are used up to several hundred litres capacity.

Another arrangement of water bottles is in the form of a so-called *rosette sampler* (Pl. 7). In this, twelve to twenty water bottles are mounted in a single frame which is attached to the end of the oceanographic wire. This has an electrical conductor incorporated and the bottles can be closed when desired by electrical command from on deck. This rosette arrangement is generally used in conjunction with a CTD sensor head with deck read-out so that water samples can be obtained to check the CTD or to obtain confirmation of interesting features in the water property profiles.

6.242 Density measurement

The standard laboratory method, using a weighing bottle, to determine density is not practical at sea because of the motion of the ship, and it is too slow for routine use on shore. The method of weighing a quartz sinker immersed in the water sample has been used by some laboratories on shore. The simple hydrometer is not to be despised for coastal or inshore work where large variations occur, particularly in the surface layers, and high accuracy is not required. Sets of three hydrometers are available to cover the range from 1000 to 1031 kg/m^3, corresponding to 0 to 41‰ in salinity which can be estimated to about 0.2‰.

A suggestion made by Richardson for the direct measurement of density was later applied by Kremling (1972). The water sample is placed in a glass tube which is supported at one end only and to the free end of which is attached a piece of iron. The tube is caused to vibrate by passing an alternating electric current through a coil near the iron and the natural frequency of vibration of the tube is determined. This depends on the mass of the tube and contents and therefore on the density of the sea-water content which is thereby determined. In the report by Kremling the accuracy was estimated at ± 0.005 in σ_t and the values obtained agreed with those of Cox et al. (1970) in indicating that Knudsen's Tables are low by about 0.013 in σ_t. This instrument has the merit of permitting continuous measurement of density if the tube is a U-tube through which the sea-water flows. It is also possible that the method might be adapted to permit the *in situ* measurement of density for which no other practical method is yet available.

More recent measurements by Millero *et al.* (1976, 1980), using a magnetic float densimeter (Millero, 1967), estimated accurate to ± 3 in 10^6 (i.e. to ± 0.003 in σ_t), indicate that Knudsen's values for relative density are low by about 0.009 in σ_t at $35°/_{oo}$ salinity.

A variety of calculations in dynamic oceanography require also the pressure effect on density. (The full relation between ρ, T, S and p is referred to as the *equation of state* of sea-water.) Previously the measurements of relative density reported by Ekman (1908) were used but recently a new equation of state was announced by Millero *et al.* (1980, see Section 3.55) with a precision of ± 5 in 10^6 (standard deviation) over the ranges of values found in the ocean, i.e. about 0.005 kg/m³ in σ_t. The new measurements indicate that Ekman's values are up to 60 in 10^6 in error, i.e. up to 0.06 in σ_t.

Failing adequate means to measure sea-water density quickly in the field with the accuracy needed for geostrophic calculations it is usual to estimate it indirectly from salinity and temperature, using the equation of state.

6.243 Salinity measurement

The classical (Knudsen) method of measurement is to determine the chlorinity by titration with standard silver nitrate solution (e.g. Strickland and Parsons, 1972) and then to calculate the salinity from the formula given in Section 3.3. In routine use, an accuracy of $\pm 0.02°/_{oo}$ is considered reasonable, with rather better accuracy if special care is taken and replicate titrations made. A careful operator may titrate fifty samples per day. It must be remembered that this method is a volumetric one, whereas salinity is defined gravimetrically (i.e. by mass). In consequence it is necessary either to correct for deviations of the temperature of the solutions from the standard, or preferably to carry out the titrations in a temperature-controlled room. This titration method is practical but not very convenient to use on board ship.

The estimation of salinity through the electrical conductivity measured by means of an A.C. bridge has been in use by the U.S. Coast Guard for the International Ice Patrol in the western North Atlantic since about 1930. The method was not more widely used for many years because of the bulk and expense of the equipment required. This is because the conductivity is as much a function of temperature as of salinity, which necessitates thermostating the samples to ± 0.001K during measurement. However, improvements in circuits and equipment encouraged a number of laboratories to bring this method into wider use from about 1956 and an accuracy of $\pm 0.003°/_{oo}$ is obtained in routine use. This is substantially better than the titration method and makes it possible to distinguish water masses which were previously not distinguishable. One of the great advantages of the electrical salinometer is that it uses a null-balance method which is much less tiring for the operator to use than the end-point method of chemical titration. However, the variability in use of the characteristics of the platinum electrodes posed problems with the earlier electrode-type salinometers.

In 1957 Esterson of the Chesapeake Bay Institute described an electrical salinometer which avoided the electrode problem by using an inductive (electrodeless) method. Then Brown and Hamon (1961) in Australia described an inductive salinometer design which has now come into wide use. In this instrument the temperature effect is taken care of not by thermostating the sample but by measuring the temperature while the conductivity is being measured and correcting for its effect automatically in the electrical circuit. The salinity may be measured to a precision of $\pm 0.003°/_{oo}$ over the range from 32 to $39°/_{oo}$, and with a little practice an operator can measure the salinity of up to forty-five samples per hour.

A new laboratory salinometer, the Canadian "Autosal" by Guildline (Pl. 8) based on a design by Dauphinee (Dauphinee and Klein, 1977), has come into wide use. This uses a four-electrode conductance cell of small dimensions in a thermostat bath (to $\pm 0.001 K/day$) with a precision of $\pm 0.001°/_{oo}$ or better. The sea-water flows continuously from the sample bottle through a heat exchanger in the thermostat, to bring it to a specified temperature, and then through the cell. The conductance bridge is balanced semi-automatically and the *conductivity ratio* of the sample relative to that of Standard Sea-Water (see below) is displayed digitally. Salinity is then obtained from the conductivity ratio and the temperature using the Unesco/N.I.O International Oceanographic Tables (until recently) or the Practical Salinity Scale 1978 Formula or Tables referred to in Section 3.3. The circuits are such that variations of electrode surface conditions do not affect the measurement. The size of the instrument is about $60 \times 50 \times 55$ cm and it may be used on shipboard as well as in a shore laboratory.

The refractive index of sea-water is also related to salinity (and to temperature) and the interference type of refractometer has been used in the past with a claimed accuracy of $\pm 0.02°/_{oo}$.

One feature of all the above methods must be noted—they are all comparative rather than absolute. The so-called *Standard Sea-Water* is prepared (now at the Institute for Ocean Sciences in England) to an accurately known salinity and conductivity ratio by comparison with a standard potassium chloride solution as described previously (Section 3.3; ref. Lewis, 1980) so that the Standard Sea-Water has a salinity of $35.000°/_{oo}$. Samples of this Standard, sealed in glass ampoules (Pl. 8), aré used by oceanographic laboratories throughout the world to standardize the silver nitrate used for titration or the electrical conductivity salinometers which are more often used now. One advantage of this procedure is that all oceanographic laboratories use a common standard for salinity, reducing the possibility of systematic errors occurring and hence making it possible to combine data from different expeditions or surveys in the same area or world-wide.

The above methods are all laboratory methods, but *in situ* measurement of water properties has always been something to aim for. About 1948 one of the first *in situ* salinometers was developed. A variety of substances and organisms in ocean waters cause fouling of the electrodes and consequent change of calibration, and in practical instruments it is necessary to design the conductivity sensor so that the electrodes may be cleaned routinely. In the mid-1950s inductive salinometers were developed for *in situ* use (Hamon, 1955; Hamon and Brown, 1958) and a number of such instruments are now available from several manufacturers (see Baker, 1981). Because they measure Conductivity, Temperature and Depth (actually pressure) they are referred to as CTD instruments. In a CTD, a unit consisting of conductivity, temperature and pressure sensors is lowered through the water on the end of an electrical conductor cable which transmits the information to indicating and recording units on board ship. The digital transmitting units have claimed accuracies of $\pm 0.005^{\circ}/_{\circ\circ}$ (conductivity accuracy expressed as equivalent salinity accuracy), $\pm 0.005K$ and $\pm 0.15\%$ of full-scale depth, with resolutions 5 to 10 times better than these figures. The sensor head of the Guildline instrument is shown in Plate 9. It uses an electrode cell for conductivity measurement, rendering calibration simpler than for the inductive type conductivity sensor because it can be carried out in a smaller calibration space. The Neil Brown instrument built in the United States uses the same principles and has become a standard of high accuracy for *in situ* measurements. There are also somewhat less sensitive self-contained instruments (STDs) in which salinity is calculated and plotted internally as a graph of salinity versus depth as well as a temperature/depth graph. These STDs are lowered on the end of the ordinary steel cable used for bottle casts. Some instruments can also measure other water properties such as dissolved oxygen content and turbidity.

In an STD, salinity is calculated from simultaneous temperature and conductivity measurements; the slower response time of the temperature sensor than of the conductivity sensor often results in large "spikes" in the salinity record. For CTDs with which salinity is computed subsequently from the temperature and conductivity records it is possible to apply some compensation for the different response times and reduce the spiking.

The above instruments are all designed for repeated use and are quite expensive. Another class of instruments, to measure temperature, salinity or other properties, which are less precise but also less expensive and are considered expendable will be described shortly.

6.244 *Temperature measurement*

For measuring the temperature of a surface bucket-sample, an ordinary mercury-in-glass thermometer is generally used, taking care not to expose the bucket to the sun (heating) or to the evaporating influence of the wind (cooling).

Another special method for determining the sea-surface temperature makes use of Stefan's Law that the rate of emission of heat radiation from an object, in this case the sea surface, is proportional to the fourth power of its absolute temperature (see Chapter 5). The radiation is measured by a radiation bolometer which uses a small thermistor as the detecting element. The electrical resistance of the thermistor depends on its temperature which depends on the amount of heat radiation falling on it from the sea. In practice the temperature of the sea is not measured absolutely but is compared with that of a constant temperature enclosure by placing the thermistor at the focus of a parabolic mirror which is wobbled rapidly so as to look alternately at the sea and at the enclosure. This gives rise to an alternating current proportional to the difference between the two temperatures. This *radiation thermometer* has its chief value in determining the sea temperature from an aircraft. In this application it can be used to examine a considerable area of sea in a short time in order to get a nearly simultaneous picture. Strictly speaking, since it operates at long wavelengths (see Chapter 5), it measures the temperature of the surface skin, a fraction of a millimetre thick, of the sea. However, in the presence of wind mixing, it is probable that this does not differ very much from the bulk temperature of the upper mixed layer. A serious source of error with the airborne radiation thermometer is the variability in signal due to the absorption by water drops in the atmosphere between the sea surface and aircraft. This requires the aircraft to fly at as low an altitude as practicable, i.e. hundreds of metres rather than thousands.

For measuring subsurface temperatures the basic instrument has been the *protected reversing thermometer* (Pl. 6) developed especially for oceanographic use. It is a mercury-in-glass thermometer which is attached to a water sampling bottle. When the latter is closed to collect the sample the thermometer is inverted and, as a result of its construction, the mercury "breaks" at a particular point and runs down to the other end of the capillary to record the temperature *in situ* at the depth of reversal. The break occurs in the capillary stem above the bulb at a point where a short side-arm is piaced. It is really rather surprising that the mercury should break as consistently as it does—to better than ±0.01K in a good thermometer in laboratory tests. After the thermometer has been reversed it becomes almost insensitive to subsequent changes of temperature and it is read when it is brought back on deck. This insensitivity subsequent to reversal is necessary because the surface temperature is usually higher than the deep-water temperature and, as it was brought back to the surface, an ordinary thermometer would warm up and "forget" the deep-water temperature. After corrections for scale errors and for the small change in reading due to any difference between the *in situ* temperature and that on deck, the reversing thermometer yields the water temperature to an accuracy of about ±0.02K in routine use.

The "protected" part of its name arises because the thermometer is enclosed in a glass outer case to protect it from the pressure of the water.

One way to determine the depth of a sampling bottle is to use an *unprotected reversing thermometer* (Pl. 6) together with a protected one. The unprotected thermometer has a hole in its glass outer case; as a result the water pressure compresses the glass of the bulb and causes this thermometer to indicate a higher apparent temperature than the protected one. The difference in reading between the two thermometers is a measure of the compression of the glass, which depends on its known compressibility and upon pressure, i.e. upon depth. A pair of thermometers, one protected and one unprotected, therefore serves to measure both the temperature *in situ* and the depth, the latter to about ±0.5% or to ±5 m whichever is the greater.

Another widely used instrument was the *bathythermograph* in which a liquid-in-metal thermometer causes a metal point to move in one direction over a smoked or gold plated glass slide which is itself moved at right angles to this direction by a pressure sensitive bellows. The instrument is lowered to its permitted limit in the water (60, 140 or 270 m) and then brought back. Since pressure is directly related to depth, the line scratched on the slide forms a graph of temperature against depth. It is read against a calibration grid to an accuracy of ±0.2K and ±2m if well calibrated. The great advantage of the bathythermograph is that, although it is less accurate that the reversing thermometer, it gives a continuous trace of temperature against depth instead of only the values at spot depths given by those thermometers. Although still in use, the mechanical bathythermograph has, to a large extent, been superseded by an electronic expendable instrument described in the next section.

In the CTD instruments described previously there are generally two thermometers. One is used to determine the temperature as a function of depth as a characteristic of ocean water masses while the other is used in the circuits for the calculation of salinity from conductivity. Platinum or copper resistance thermometers are generally used for both purposes.

Thermistor "chains" consisting of a cable with a number of thermistor elements at intervals are sometimes moored along with current meters to record the temperature at a number of points in the water column. A "data logger" samples each thermistor sequentially at intervals and records temperatures as a function of time.

6.25 Expendable instruments

In wide use now is the *expendable bathythermograph* (XBT) which uses a thermistor as temperature-sensitive element (Pl. 10). The thermistor is in a small streamlined weighted casing which is simply dropped over the ship's side. It is connected by a fine wire, on special free-unwinding spools, to a recorder on the ship which traces the temperature of the water in a graphical plot against depth. The latter is not sensed directly but is estimated from the time elapsed since release, using the known rate of sink of the freely falling thermistor casing. This casing is relatively inexpensive and is not recovered. These XBTs are available for depth ranges from 200 to 1800 m, and can be used from ships underway and have even been used from circling aircraft. They can also be dropped from aircraft (AXBT) in a small buoy which contains a radio transmitter to send the temperature/depth information (to 300 to 800 m) to the aircraft which is continuing its flight.

Expendable instruments have provided the oceanographer with simple tools for rapid sampling. This has proved important for synoptic sampling from multi-ship or aircraft surveys and has led to wider use of *ships of opportunity*. These are regular passenger or cargo vessels sailing through an area of interest to the oceanographer and from which simple measurements may be made without stopping or interfering with the ship's normal passage routine. They have ranged from small coastal vessels to super tankers. In an effort to extend such technology to other important parameters, recent developments have produced an expendable velocimeter (speed of sound) and an expendable current profiler, using the electro-magnetic principle. An expendable CTD (XCTD) is also available but needs further development to be truly reliable. Plans also call for the construction of an expendable microstructure velocity profiler to study turbulence on the centimetre scale. These more exotic expendables are all considerably more expensive than the XBT and therefore have been less widely used.

6.26 Radiation measurement

Direct measurements of Q_s are made with a pyranometer. The sensing element of the Eppley pyranometer consists of two flat plates of copper, one painted with a flat black paint and the other whitened with magnesium oxide. The two plates are placed horizontally with a clear view of the sun and sky and are shielded from draughts by a clear hemispherical cover. The black paint absorbs all radiant energy, short- and long-wave, falling upon it and is thereby heated above the surrounding temperature. The white plate reflects practically all of the energy between 0.3 and 5 μm (short-wave radiation) but absorbs all long-wave energy. The white plate is consequently heated less than the black one and the difference in temperature between them is a measure of the short-wave radiation (Q_s) falling on a horizontal surface in the locality of the instrument. The difference in temperature is measured by connecting the "hot" junctions of a group of thermocouples to the black plate and the "cold" junctions to the white plate. The difference in temperature gives rise to a thermoelectric EMF which is measured by recording galvanometer. The instrument is calibrated by exposing it to a standard source of energy, such as a standard electric filament lamp.

The downward directed component of the long-wave radiation term Q_b is measured with a radiometer. The Gier and Dunkle instrument consists of two horizontal plates of black material separated by a layer of material of known heat conductivity. The upper sheet of black material absorbs all the radiation falling upon it from above and is thereby heated above the temperature of the lower sheet which is screened from radiation from below by a sheet of polished metal. The difference in temperature between the upper and lower sheets is measured by thermocouples and is a measure of the rate at which the sum total of long- and short-wave energy is coming down from above. To determine the value of the long-wave component itself it is necessary to subtract the short-wave radiation rate as measured with a pyranometer. An alternative procedure is to omit the polished metal screen from below the black horizontal plate and arrange the instrument so that the upper plate "looks at" the atmosphere and the lower plate "looks at" the sea below. In this "net radiometer" arrangement the difference in temperature between the upper and lower plates is a measure of the net amount of radiant energy reaching a horizontal surface, i.e. it is a direct measure of $(Q_s - Q_b)$.

To determine the transmission of visible light through the water the simplest device is the *Secchi-disc*, a white plate about 30 cm in diameter fastened to hang horizontally on the end of a rope marked in metres. The disc is lowered into the sea and the depth at which it is lost to sight is noted. This depth decreases as the vertical attenuation coefficient of the sea-water increases. In very clear water the depth may be over 50 m, in coastal waters 10 to 2 m, and in some river estuaries less than 1 m. The Secchi-disc is only a semi-quantitative device but being simple it is often used. When one first starts using the device it is rather surprising to find that after a very little practice it is possible to obtain consistent readings to better than 10 % with little variation from individual to individual. The Secchi-disc gives an estimate of the vertical attenuation coefficient k between the surface and the Secchi-disc depth D. Where the depth D is greater than a few metres it has been shown that for blue light $k = (1.6 \pm 0.6)/D$.

A more quantitative instrument for determining the (beam) attenuation coefficient is called a *transparency meter* or a *turbidity meter*. In this, the sensing head which is lowered into the water has a lamp and a photoelectric cell in separate watertight housings mounted 0.5 to 2 m apart and opposite to each other. The principle of the instrument is that the current from the photocell is a measure of the amount of light falling upon it. The lamp has a lens and collimator to ensure a parallel beam of light which passes through the water and falls upon the photocell. The light output is kept constant by regulating the current through the lamp. The light beam is usually arranged horizontally because the absorption characteristics in the sea are often horizontally stratified. Also the photocell must not face upward or it would pick up direct sunlight which would give too high an apparent transmission by the water. The photocell is connected by an electrical cable to a microammeter on deck. The current from the photocell is recorded when the sensing head is on deck with the light passing through air whose attenuation is taken to be zero. In the water the photocurrent decreases due to the attenuation of light by the water. The decrease from the air path value is a measure of the attenuation by the water. If the initial light intensity is I_o and that in water after a path length x is I_x, the beam attenuation coefficient C may be obtained from the formula $I_x = I_o \cdot \exp(-C \cdot x)$ or in logarithms to base 10 as $C = 2.3 \, (\log I_o - \log I_x)/x$. The length x used in these formulae is the length of the water path between the lamp and photocell in metres (usually from 0.1 to 2 m). The instrument is lowered in stages and the attenuation coefficient measured at a series of depths to give a vertical profile for this water property. The attenuation is often associated with particulate material present in the water independently of its temperature and salinity and therefore may be useful as an independent tracer of water masses. The attenuation coefficient C, or the "turbidity" as it is sometimes called, has been used to distinguish water masses both in the open ocean and in coastal fjords.

It must be noted that the parallel beam attenuation coefficient C is not identical with the vertical attenuation coefficient k (Section 3.8) where light at a point in the sea may be arriving from a variety of directions. Both C and k represent loss of light both by absorption (i.e. conversion to heat or chemical energy) and of scattering (i.e. change of direction while remaining radiant energy). For both beam and vertical attenuation, absorption is effective in reducing the light intensity with distance. However, scattering is more important for beam attenuation because *any* scattering represents loss of energy from the beam whereas for vertical attenuation we are concerned with energy arriving at a point in the sea where light is travelling in a variety of directions and a significant part of the energy arriving there will have been scattered from light travelling in other directions. If we express the beam attenuation coefficient C as the sum of absorption (a) and scattering (b), i.e. $C = a + b$, then theory and observation suggest that roughly $k = 1.4a + 0.03b$, i.e. scattering contributes relatively little to k compared with absorption (e.g. see Jerlov, 1976).

For some purposes it is desired to know the amount of daylight which reaches a particular depth. This is done by mounting a photocell behind a filter in a watertight housing to face upward. A microammeter, connected to the cell, is read first with the cell on deck facing upward to give a measure of the sun and sky light falling on the sea surface in the spectral range passed by the filter. The cell is then lowered to the depth of interest and the photocurrent again measured. The ratio of the photocurrent with the cell in the water to that on deck is then a measure of the ratio of light intensity at depth to that at the surface. If the photocell indications have been calibrated against a standard light intensity meter then the actual light intensity at depth will be known. It is usual to measure the light intensity at a series of depths below the surface in order to obtain a profile of light intensity as a function of depth. It is usual to mount an opal glass or hemisphere in front of the photocell so that it will collect light from all directions above its plane.

In the above instruments, the Weston type of photovoltaic cell is often used because its response can be made substantially linear with light intensity at a particular wavelength, because it responds over the full range of visible light and because it does not require a separate source of power. However, the Weston cell has a limited sensitivity and it may be inadequate if measurements are required of the *in situ* light intensity at considerable depths. In this case it is necessary to use the more sensitive photomultiplier type of cell. The disadvantage of this type is that it requires a high-voltage supply and precautions have to be taken to avoid electrical leakage when working in sea-water.

6.27 Platforms

6.271 Sea- and airborne

The majority of the oceanographic measurements in the past were made from ships, with contributions from piers, etc. (for surface properties, tides and waves), but other *platforms* for carrying men and instruments are coming into use. The pier measurements are being supplemented by observations from oil drilling platforms at sea, and the development of these facilities has depended very much on knowledge and forecasts of currents and waves which may affect the structures.

The techniques for supporting current meters, etc., from buoys have already been discussed. Considerable skill is needed for the design of the instruments and the techniques for placing the buoys and instruments in position and recovering them. One hundred percent acquisition of data is rarely achieved but techniques have improved from the 10 % or so obtained 20 years ago to 80 to 90 % or even better nowadays.

Ground-effect vehicles ("Hovercraft") are found to be very effective in shallow waters such as estuaries and mud-flat regions and in ice-infested areas. Observation of the sea-surface temperature is now done routinely from aircraft with the airborne radiation thermometer (ART) described previously. With the speed of the aircraft platform it is possible to cover considerable areas quickly with that instrument and AXBTs. Sequential photography from aircraft can be used to track floating objects such as icebergs or dye patches, and can use natural colour differences, such as silt from a river, to determine flow patterns.

6.272 Remote sensing; satellites

The most extensive horizontal coverage is obtained from satellites, those in polar orbits providing world-wide observations with repetitive cover of the same surface location twice per day, while geosynchronous satellites, which remain stationary relative to the earth over the equator, provide each half-hour an image of nearly one-half of the earth's surface at once. The satellites in polar orbit are at heights of 500 to 1500 km while the geosynchronous ones must be at about 36,000 km above the earth's surface. Manned spacecraft observations have been made from heights as low as 180 km.

The sensors used include photographic and television cameras as well as optical scanners for observations in the visible (0.35 to 0.7 μm), radiometers for the infra-red (particularly the 8 to 14 μm atmospheric "window") and for micro-waves (about 1 cm) which can "see" through cloud which blinds the infra-red sensors and, most recently, imaging and altimetric radars. The spatial resolution of the sensors is either 7 km or 1 km diameter areas on the sea for the infra-red (to 0.1 km in special cases) and to 1 km or less for visible light, while the new radar images have a resolution to 25 m.

The information available to oceanographers which is obtained or anticipated in the near future from satellite observations includes:

(a) Sea-surface temperature (SST) routinely determined to $\pm K$ with $\pm 0.5K$ expected soon. (It should be noted that radiometric SSTs are necessarily restricted to a centimetre thick layer at the surface.)

A good example of the complex structure revealed by satellite infra-red imagery is shown in Pl. 11 where, in an image from October 2, 1980, the west coast of North America is cloud free from the northern tip of Vancouver Island (top, left of centre) to northern California (bottom right). For the water, white represents the coldest temperatures and black the warmest. The infra-red SST reveals the presence of cold water along the coast extending westward into the North Pacific. Although there is a general trend offshore from warm to cold when moving north, the pattern is quite complicated with eddies and meanders dominating the scene. Features are generally larger at the lower latitudes and smaller-scale features dominate the boundaries between warm and cold water. This truly synoptic view of the ocean indicates the complex spatial variability of features being studied by oceanographers.

(b) Radiation components to about $\pm 3 \, \text{W/m}^2$.

(c) Rainfall estimates over the open ocean may be inferred from the distribution of highly reflective cloud.

(d) Location of upwelling from temperature and from colour (chlorophyll production by phytoplankton) and estimates of primary production from chlorophyll estimates).

(e) Currents estimated from silt patterns near coasts, by tracking free-drifting buoys in the open ocean and, anticipated in the future, from the slope of the sea surface determined from altimeter measurements relative to the known satellite orbit and the shape of the geoid.

(f) Ice cover is routinely observed, cloud permitting, and icebergs tracked from successive satellite passes, with resolution down to 100 m size now and 25 m size expected in the near future with the new radar equipment which will also permit observations of ice through cloud.

(g) Mean wave-height measurements to ± 1 m or 25 % of the actual height are anticipated from the altimeter observations, while measurements of tides and tsunami in the open ocean are expected.

(h) Surface wind speeds to ± 2 m/s and $\pm 20°$ direction are anticipated from observations of microwave scattering from the sea surface.

Many other items of oceanographic information may be inferred from satellite observations and while some may not be of high precision the wide coverage in space and time may compensate to some extent and render the information valuable.

Another important application of satellites is for data acquisition from surface instruments. A satellite at 500 km altitude seems very remote from the sea but in fact it is often much closer to a mid-ocean buoy than is the nearest shore station. Techniques for transmitting data by microwave link from buoy to satellite to shore station are available, using geostationary satellites. As has been pointed out, polar orbiting satellites are also instrumented to track and interrogate drogued, drifting buoys.

The satellites from which oceanographic data have been obtained until 1977 were primarily designed for meteorological or land resource studies. They include the TIROS (Television Infra-Red Observational Satellite) series starting in 1960 and the *Nimbus* series (for cloud studies) from 1964 using visible and infra-red sensors, followed by the ESSA and NOAA series from 1966 to 1975, all of these being in polar orbit. In geosynchronous (geostationary) orbit were or are the ATS (Advanced Technology Satellite, 1966 on) and GOES (Geosynchronous Operational Environmental Satellite, 1974 on). Later ones in polar orbit are ERTS (Earth Resources Technology Satellite, 1972) and *Landsat* (1974 on). The first satellite dedicated to oceanographic observations was *Seasat*, launched in 1978, which carried radiometers for visible, infra-red and microwaves, radar scatterometers, imaging radar and altimeter in a near polar orbit at 800 km height and providing cover of 95 % of the oceans every 36 hours. Unfortunately it ceased to function after 3 months but gave extremely valuable information in that time and promises excellent potential for future satellites in the series. Modern polar orbiters are two TIROS-N satellites and similar spacecraft called NOAA-6 with NOAA-7.

Finally, the U.S. Navy Navigation Satellite System should be mentioned. Several satellites are in polar orbits which are checked frequently from the ground. The satellite emits radio waves which are picked up by a receiver on the ship. The relative motion of satellite and ship gives rise to a Doppler shift of frequency of the received signals and from a record of these and the orbit information (also transmitted by the satellite) a small computer can, in a few minutes, determine the ship's position to within a fraction of a mile, day or night, clear or cloudy, anywhere on the oceans. This is a great improvment on earlier techniques. A more advanced system, *Navstar*, is planned for the near future with more satellites, greater accuracy and three-dimensional capability, i.e. it will also give height.

Some of the satellite data are used immediately, e.g. the cloud pictures which are used for daily forecasts and to illustrate weather reports on television, and the navigation system is in continuous use; other data are accumulated over long periods for climatological studies. Examples of this anticipated from *Seasat* data were to study the relations between SST anomalies in the North Pacific and anomalous winter weather on the North American continent, and for studies in connection with *El Niño* (Chapter 7), both of which require data from larger areas than can possibly be studied by ships. Radiation data for global radiation budget studies are also accumulated.

6.28 Age of ocean water

The term *age* as applied to ocean water means the time since the water mass was last at the surface and in contact with the atmosphere. The importance of this characteristic is that it gives some indication of the rate of overturn of ocean water. This is of interest in connection with the rate of replenishment of nutrients in the upper layers and of the use of the ocean basins as places to dump noxious material, particularly radioactive waste. If the average time of overturn is much less than the half-life of such materials it would be dangerous to dump them in the ocean because they would be brought to the surface while still active and might be picked up by fish and so conveyed back to man. We could easily calculate the age of ocean water if we knew the speed of travel of deep currents but this is exactly what we know so little about. For this reason, attempts have been made to develop other methods for estimating the age, and some of these will be described.

One method, devised by Worthington (1954), was to use the rate of consumption of dissolved oxygen. He observed that at depths of about 2500 m (in the North Atlantic Deep Water) the average oxygen content decreased by 0.3 mL/L between 1930 and 1950. He attributed the decrease to steady consumption of oxygen by chemical combination with detritus, the rate being 0.3 mL/L in 20 years = 0.015 mL/L per year. Assuming that the water had been saturated with oxygen when it was last at the surface it would have decreased in oxygen content from its saturation value of 7.6 mL/L at the surface to the 5.8 mL/L observed in 1930 in $(7.6 - 5.8)$ mL/L ÷ 0.015 mL/L per year = 120 years (assuming the consumption rate to have been constant). This implied that the water was last at the surface in 1810. This was in a period of very cold climate and Worthington suggested that much of the present North Atlantic Deep Water may have been formed cataclysmically at that time, and that relatively little had been added since. This use of the oxygen-consumption rate to measure the age has been criticized on the grounds that the accuracy of measurement of dissolved oxygen may not be sufficient for the figure of 0.015 mL/L per year to be very reliable, and on the grounds that the assumption of constant oxygen consumption may be unreal. However, the method is a legitimate attempt to wrest information from the distribution of properties.

Another method, first reported for ocean water by Kulp *et al.* (1952), was to use the decay rate of ^{14}C in deep waters. It was assumed that the atmosphere at the sea surface was the only source of ^{14}C to ocean waters. Away from the surface the ^{14}C content would not be replenished and it would decay with its half-life of about 5600 years. The early measurements suggested an age of the order of 2000 years for water at 2000 to 5000 m in the North Atlantic, but this was subsequently shown to be too high on account of contamination in the chemical processing. The most recent estimates for replacement or residence times in the Atlantic, made by Broecker (1979), based on GEOSECS studies (described later) suggest 100 years for North Atlantic Deep Water in the western basin of the Atlantic and 550 years for the deep world ocean (deeper than 1500 m). These estimates are based on rates of production of North Atlantic Deep Water of 30×10^6 m^3/s and of Antarctic Bottom Water of 20×10^6 m^3/s. (These are names for distinct water masses to be described in Chapter 7.)

Measurements by the ^{14}C method in the South Pacific have indicated ages of some 1100 to 1900 years for the Deep Water with a suggestion that the residence time (volume of water in the region divided by the rate of inflow or outflow) is about one-half of this.

Apart from the experimental uncertainty in the method of measuring the ^{14}C activity in a sample of sea-water (a 200-litre sample is needed) there are complications associated with exchange between living and dead matter in the sea and even greater ones resulting from changes in ^{14}C concentration in the atmosphere in the present century. The rapid increase in the use of fossil fuels since about 1900 has reduced the ^{14}C concentration in the atmosphere because the CO_2 from the fuels contains mostly the stable isotope ^{12}C, the fuels having been away from the atmosphere for so long. In the opposite direction, since 1954 the effect of nuclear explosions in the atmosphere has been to cause a huge increase in ^{14}C.

Measurement of the ^{90}Sr content of ocean water has revealed significant amounts at depths to 1000 m. As the only source of this isotope is presumed to be the residue from atom bombs, starting in 1954, this indicates that the rate of vertical mixing in the upper waters may be quite rapid.

Another radioactive tracer which is being used is tritium (^{3}H) with a half-life of about 12 years. It occurs in the upper layers of the ocean at concentrations of the order of only one tritium atom for 10^{17} or 10^{18} ordinary hydrogen atoms (^{1}H) but with new techniques it can be measured quantitatively for age determinations of sea-water which have been routinely measured in the systematic redetermination of the distribution of the main elements in the world oceans in the GEOSECS (Geochemical Ocean Sections Study) which was a multi-national, multi-institutional study of the main oceans whose objective was the study of the geochemical properties of the oceans with respect to large-scale circulation problems. It was a part of the International Decade of Ocean Exploration.

6.3 Graphical Presentation of Data

6.31 Variation in space; profiles and sections

6.311 Vertical direction

The synoptic oceanographer cannot rely on the distribution of one property alone to determine the full story of the ocean circulation. The discussion in Chapter 7 of the four sections of Fig. 7.9 will make this apparent. He must use as much independent information as he can obtain, but he must be careful that his interpretation of the circulation is consistent with all the property distributions.

The distribution of properties with depth is best shown on the temperature/depth or salinity/depth *profiles* (e.g. Figs. 4.5 and 4.11) which are usually drawn as the first stage in examining oceanographic data. When taking oceanographic stations at sea the oceanographer tries to make observations at standard depths to facilitate comparison between stations. However, the drift of the ship and effects of currents generally cause the oceanographic wire to be at an angle to the vertical and the observations often come out at other than the desired standard depths. If this is the case it is possible to interpolate on the profiles to get values at the required standard depths. A more sophisticated way to do this is to feed the "raw" (observed) data into a computer which has been programmed to interpolate between the observed points according to some standard procedure and to give out the values at the required depths. The computer is usually programmed to do more than this. Not only does it interpolate for the desired depths but prepares, as output, a complete master sheet with observed and interpolated data, station position, date, meteorological observations, etc., in a standard format ready to be duplicated for the *data report* which is one of the first products from an oceanographic cruise. This data report is simply a list of the numerical values for the observations made at sea, corrected for instrument errors, etc., but without any attempt being made to interpret them.

Vertical profiles are one way of displaying data from a single oceanographic station. For a number of stations, and when other parameters enter, such as time or a range of geographical positions, it is necessary to use further displays. For example, in Fig. 4.6(a) are presented vertical profiles of temperature at a fixed station for a number of months. The variation of temperature as a function of time is then shown in two ways in Figs. 4.6(b) and 4.6(c). These plots were discussed in Chapter 4.

Vertical sections, such as those of Figs. 7.6, 7.9, etc., are used to display an aspect of the geographic distribution of properties in the vertical direction, e.g. along a line of stations, usually a straight or substantially straight line. Such sections, if well chosen, can be useful in showing the path followed by a water body and the changes which take place along such a path. The "core method" developed by Wüst (Section 6.342) uses this display.

6.312 Horizontal direction

The horizontal *sections* (*maps*) of properties such as those of Figs. 4.3, 4.4 and 4.9 are most useful in displaying the geographical distributions of individual properties. They are not generally used to show the interdependence of the properties because it is difficult to show more than two properties simultaneously on one diagram, even using multiple colours. Even with only two properties, the relations between them cannot be appreciated as well, for instance, as on the characteristic diagrams described later in this chapter.

6.32 Variation in time

6.321 Time-series plots

To display the variation of properties as a function of the fourth parameter, time, we may initially use plots which are basically similar to those already described for spatial variations. The commonest is the *property time-series* in which the property value is plotted along the y-axis and time along the x-axis, e.g. Fig. 6.7(a). To show the relations between the time variations of several properties, e.g. temperature, salinity, such time series plots may be arranged one below another down a page with the time scales aligned vertically. A variation is a *time-series of profiles* in which a temporal series of profiles is plotted across a page at positions along a time axis corresponding to their time of observation, e.g. Fig. 6.7(b). (The location of such profiles along the time axis would probably be made relative to the deep end of the profile because temporal variations are likely to be least there.) If the successive profiles are markedly different, they may be plotted on a single vertical profile diagram as

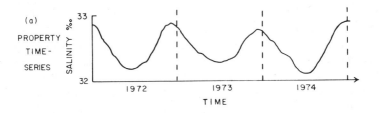

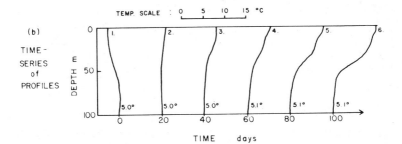

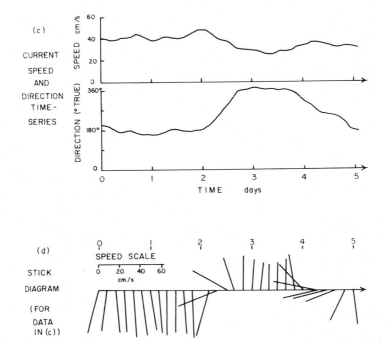

FIG. 6.7.   *Examples of time-series plots: (a) property/time, (b) time-series of profiles, (c) current speed and direction, (d) stick diagram for data of (c) above.*

in Fig. 4.6(a). The simultaneous display of two properties with time, e.g. the (TS*t*) plot, is described in Section 6.343.

A *time section* plot would have depth along the *y*-axis and time along the *x*-axis with numerical values of the property initially entered on the plot and then contoured as in the isotherm plot of Fig. 4.6(b).

A more sophisticated procedure is to prepare a *spectral analysis* of property fluctuations with time to determine any periodic characteristics, e.g. diurnal, tidal. For discussions of spectral analysis techniques the reader is referred to more advanced texts (e.g. see Jenkins and Watts, 1969).

6.322 Current data plots

Current, being a vector, is more difficult to present because it must be described by two quantities, magnitude (i.e. speed) and direction, in contrast with the scalar quantities considered so far (e.g. temperature, salinity) which have magnitude only.

For observations at one point, a pair of time-series plots of speed and direction respectively may be presented, one above the other, with time scales aligned vertically (Fig. 6.7(c)). Data from a string of current meters at one mooring can then be presented with the speed and direction pairs of plots one above another. Related data such as wind speed and direction or tidal rise and fall may be displayed on parallel time-series plots.

Another display method for vectors is the *stick plot* in which velocity is represented in speed and direction by a line drawn to scale out from a time axis. A series of velocities is then represented by a series of "sticks" drawn in succession from the time axis at their time of observation as in Fig. 6.7(d). Again, currents from a string of meters on a mooring can be plotted one above another down a page to give an idea of the correlation between depths.

An alternative procedure is to represent the velocities as time-series of two *rectangular components* either (a) east-west and north-south or (b) parallel and perpendicular to a shore-line or the centre-line of a passage.

Vertical sections of current components perpendicular and parallel to the geographic direction of a section may be prepared by entering numerical values of the components on the sections at their measurement positions and then contouring the data. The north—south equatorial section of east—west current components in Fig. 7.27(a) showing the Equatorial Undercurrent is an example.

Horizontal maps of current observations are often used, as in Figs. 7.4 and 7.24, to show compilations of current observations or to show the current pattern derived from the dynamic topography of a surface as in Figs. 6.5 and 6.6.

Time-series current data are usually subjected to spectral analysis to determine the periodic components, e.g. tidal or due to internal waves, leaving the residual non-periodic components.

6.33 Isentropic analysis

Although it will not be used in this introduction, a student who goes further in descriptive oceanography will soon come across another technique, analogous to horizontal sections, which was introduced for oceanographic use by Montgomery (1938) and referred to as *isentropic analysis*. He argued that the flow of ocean waters may be expected to occur most easily along surfaces of constant entropy. Because of the complex thermodynamic nature of sea-water it is difficult to determine surfaces of constant entropy as such and Montgomery presented arguments to show that surfaces of constant potential density (σ_θ) would be a close approximation, while for the upper layer of the ocean the more easily determined surfaces of constant σ_t (which we will call isopycnal surfaces) would be adequate. He argued, therefore, that flow should occur preferentially along isopycnal surfaces and that mixing between water masses would occur more freely parallel to these surfaces (called "lateral" mixing) than across them (which would be hindered by stability). A comparison of Figs. 7.9 and 7.34 shows that flows indicated by cores (see Section 6.342) of salinity and of oxygen are approximately along σ_θ surfaces in the former or σ_t surfaces in the latter. (In the deep water of the Atlantic it is better to use σ_4 surfaces as discussed in Section 7.352.) The procedure is first to plot T–S, T–O_2, etc., diagrams (see the next section) for each station in the region under study. Then for each selected σ_θ or σ_t the salinity and temperature at each station are read off the T–S diagram, and the oxygen value for the temperature is read off the T–O_2 diagram. The salinity and oxygen values are plotted on charts of the region and the data contoured. The tongues of extreme values of properties then show flow directions, and mixing is indicated by changes in values of the properties along the flows. As an example, Montgomery applied this method to study the flow patterns in the southern North Atlantic, and showed that in the upper few hundred metres, the flow patterns on σ_t surfaces as delineated by salinity and dissolved oxygen distributions were generally simpler and more continuous than on surfaces of constant geometrical depth. (The σ_t surfaces are generally not horizontal over substantial distances and so surfaces of constant geometrical depth cut through them and therefore through the flow patterns.) This method of presenting property distributions on constant σ_t (or constant $\Delta_{S,T}$) surfaces has been widely used, although it is now recognized that there are situations in which mixing across σ_t surfaces does occur. Notice that as σ_t is determined principally by temperature in the upper layers of the ocean, a plot of temperature on σ_t surfaces is redundant.

6.34 Characteristic diagrams

6.341 Two characteristics, e.g. $T-S$, $T-O_2$

Although one can plot separate vertical profiles and sections of temperature, salinity, dissolved oxygen, etc., one must be careful not to consider the distributions of these properties to be independent. Physically the properties are independent of each other in the sense that in a sample of sea-water it is possible to alter the value of any one without altering the others. (Note that density is not included in the above list because it is dependent on temperature and salinity.) However, observations indicate that in the oceans the properties do not occur widely in all possible combinations. The reason for this is that most ocean water masses acquire their characteristics at the surface of the sea in particular localities. The water properties are determined there by the local climate, and when the water sinks along density surfaces it carries these properties with it. The result is that instead of all possible combinations of temperature, salinity, oxygen, etc., occurring in the ocean, we find that only a limited number of particular combinations occur, and these occur in different regions. In consequence, we can often recognize a water mass by its characteristic combination of water properties.

To show these combinations, it is usual to plot *characteristic diagrams* of the water properties. Helland-Hansen (1916) first suggested this technique by plotting temperature against salinity (T–S plots) for individual oceanographic stations, in addition to the separate plots of temperature, salinity, etc., against depth. To prepare a T–S plot for a station, a point is plotted for the temperature/salinity combination for each bottle depth, or for a selection of depths from CTD data, and the points then joined in order of depth by a smooth (but generally not straight) line which is called the $T-S$ *curve* for that station. (Figures 7.13 and 7.14 contain smoothed, average examples for the main oceans; they will be discussed in Chapter 7.) Other characteristic diagrams such as T–O_2, S–O_2, O_2–phosphate, etc., may also be plotted.

It is not necessary to plot density as a property on a characteristic diagram because it depends on temperature and salinity and therefore appears automatically on the T–S diagram. Each point on the diagram corresponds to a particular combination of temperature and salinity and therefore to a particular density. However, the same density may be attained by different combinations of temperature and salinity, and these combinations lie on a smooth curve on the T–S plotting sheet and can be drawn in as shown by the full lines on Fig. 3.1 or 6.8.

When discussing the T–S diagram, a water body whose properties are represented by a point is called a *water type*, while one represented by a line is called a *water mass*. These are ideal definitions. In practice the points on a T–S

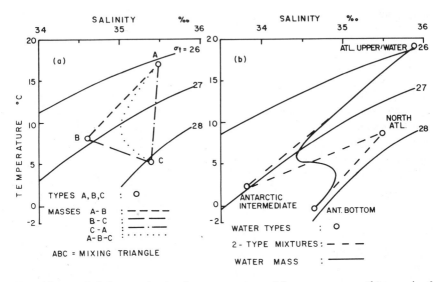

FIG. 6.8.   (a) T–S diagram showing three water types and four water masses, (b) example of realistic T–S water mass diagram (Atlantic) and basic water types.

diagram representing a realistic water type would have some scatter about the ideal point while the points representing a water mass would have some scatter about an ideal line. Some examples of these scatters will be given when discussing Pacific and Atlantic Ocean water masses in Chapter 7. Climatic processes tend to form water types at the surface; a water mass results from the mixing of two or more water types. For instance, in Fig. 6.8(a) the small circles A, B and C represent arbitrary water types while the straight dashed lines AB, BC and CA represent water masses made up of mixtures of A and B, B and C, and C and A respectively. The curved dotted line represents one example of a water mass made up of the three types A, B and C. Lines within the *mixing triangle* ABC represent all possible masses from types A, B and C. This interpretation implies that temperature and salinity are conservative quantities in the ocean, i.e. that no processes exist for generating or removing heat or salt. This is true within the body of the ocean, but near the surface it is not so. Here, the sun may heat the water or evaporation cool it, rain may decrease the salinity or evaporation increase it. Therefore in plotting the T–S diagram it is usual to regard the points corresponding to shallow depths within the influence of the surface effects as less conservative, or even to omit them altogether.

The T–S diagram turns out to be a powerful tool for the study of ocean waters (e.g. see Mameyev, 1975). The shape of the T–S curve is often

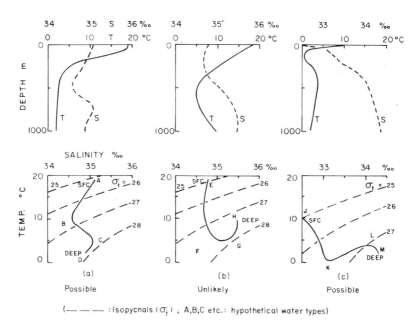

(— — — — : Isopycnals (σ_t) , A,B,C etc.: hypothetical water types)

FIG. 6.9.   *Temperature and salinity profiles and corresponding T–S diagrams.*

characteristic of water from a particular locality in the ocean, and individual features of the curve may indicate mixtures of different types of water. For example, the Atlantic water masses could be represented to a close approximation as a mixture of four types as shown schematically in Fig. 6.8(b) which may be compared to mean T–S curves for that ocean in Fig. 7.14.

Salinity minima or maxima such as B and C in Fig. 6.9(a) are common on T–S diagrams but, except for the surface and bottom values, temperature minima or maxima are uncommon. The reason is that except in polar or coastal regions, density depends chiefly on temperature and less on salinity. A temperature minimum, such as in Fig. 6.9(b), would imply that the water below the minimum, i.e. toward H, was less dense than that at the temperature minimum and therefore that the water between G and H was unstable. Such instabilities are found occasionally, such as in the complicated water masses at the northern side of the Gulf Stream where the Labrador Current joins it, but only over very limited depth ranges and probably only as a transient phenomenon.

On the other hand, in polar regions where the salinity and temperature are both low the density depends chiefly on the salinity. Under these conditions the isopycnals (σ_t) on the T–S diagram are more nearly parallel to the temperature axis as in Fig. 6.8(c) and a temperature minimum (K) or maximum (L) is quite possible from the surface to the deep water.

The T–S curve is also useful for checking data to determine if individual points may be in error. If a point lies well off the expected curve for the region one tends to regard it with suspicion.

The T–S diagram, however, suffers from two disadvantages. One is that it gives a poor indication of the distribution in depth of the different water properties because the depth scale along the T–S curve not linear (see Figs. 7.13, 7.14 or 7.21(b)). For this reason it is not possible to interpolate temperature or salinity values for depth on a T–S curve, except very roughly. The other disadvantage is that the T–S diagram gives no indication of the spatial distribution of water properties; this is better done on vertical or horizontal sections.

6.342 Core method

Tongue-like distributions of water properties in the vertical plane are very helpful in indicating the direction of movement of the water. Wüst (1935, 1957) developed the *core method* as a technique for determining water-flow paths in the ocean. A "core" is defined as a region where a water property reaches a maximum or minimum value within a tongue-like distribution, e.g. the salinity minimum of the Antarctic Intermediate Water or the salinity and oxygen maxima of the North Atlantic Deep Water (Fig. 7.13). A vertical section of the water property is used to locate such a core which is taken as the centre of the flow. A core gradually weakens along its length as a result of mixing with the surrounding water. A T–S curve may be plotted from the temperature/salinity values for a series of samples of water along the core itself from its start at the region of formation of the water type, usually at the surface, to its end where it can no longer be detected. The point on the diagram representing the original water type is taken to represent 100% concentration and the end point represents 0% concentration of the original type. The position of any point on the T–S line then represents the proportion of original water remaining by its proportionate distance from the starting point. Another use of the core method for tracing the flow of Atlantic Water round the Arctic Ocean is described in Chapter 7.

Care must be taken in interpreting tongues or cores in the horizontal plane as being indicative of flow direction. If the water property does not influence density significantly, e.g. dissolved oxygen or small amounts of silt, then a horizontal tongue may well indicate flow direction but a temperature tongue may not. This is because temperature has a strong influence in determining density and, as can be appreciated from the discussion of the geostrophic method (Chapter 6), flow tends to be parallel to isopycnals, i.e. along the isotherms outlining a tongue and not along the core which would be across the isotherms/isopycnals. The flow should therefore be around the tongue, not along it.

6.343 Three characteristics, e.g. T–S–V, T–S–t

An interesting development of the T–S diagram was developed by Montgomery (1958) and used by him and his colleagues (Cochrane, 1958; Pollak, 1958) to display the distribution of temperature and salinity of the ocean waters in proportion to their volume. Essentially a T–S plotting sheet was divided into a grid of squares, say 2K by 1‰. On each square was entered the volume of water whose property values lay within those of the square. For this purpose the potential temperature was used rather than the *in situ* temperature. The volume information was arrived at from observed oceanographic data in the following manner. Oceanographic stations were selected for which measurements of water properties extended from the surface to the bottom. The stations were as evenly spaced as possible and each was taken to represent a horizontal area around it so that the sum total of all the station areas was equal to the total area of the ocean being considered. At each station, the temperature and salinity observed at each depth were taken to represent the layer of water extending from half-way to the observation above to half-way to the observation below. The product of the interval of depth with the horizontal area represented by the station gave the volume of water for which those values of temperature and salinity were representative. This volume was entered in the appropriate grid square on the T–S diagram. This procedure was carried out for all the temperature/salinity values for all the stations and then the volumes in each square were totalled. The final result was the gridded volumetric T–S diagram in which each square showed the volume of water in the ocean having the temperature and salinity values of that square. It will be called a T–S–V diagram. On a typical such diagram, some squares have large volumes and many have none at all.

As an example of the result, a T–S–V diagram for the Atlantic and adjacent seas is shown as Fig. 6.10(a). This figure is slightly simplified from Montgomery's original diagram; the unit of volume is 10^5 km³. The number in each square then shows the number of these units of volume which have the temperature and salinity within the range of that square. It is seen that the water properties are by no means uniformly distributed but that there is a concentration of volume near $2°C$ and $35‰$. This North Atlantic Deep Water, with a volume of 1600×10^5 km³, is seen to comprise some 45% of the total volume of the Atlantic (3528×10^5 km³). The Antarctic Bottom Water and the Arctic Deep have similar characteristics and fall in the same square on the diagram (near $0°C$, $35‰$) with a combined volume of 366×10^5 km³

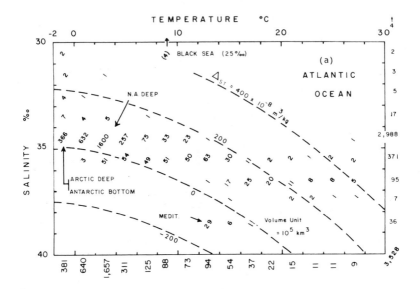

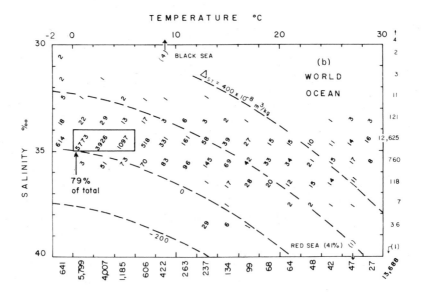

FIG. 6.10. *Temperature–salinity–volume (T–S–V) diagrams for the Atlantic and the World Ocean.*

Two other distinctive water masses are the Mediterranean Water with characteristic high salinity and temperature, and the Black Sea water with low salinity.

The numbers outside the axes of the diagram are the sums of the volume numbers in the horizontal rows or in the vertical columns respectively on the diagram. Down the right-hand side, this series of numbers shows that the great bulk (85 %) of the Atlantic water has a salinity between 34 and 35‰. In temperature, 47 % of the water is between 2° and 4°C and 76 % is between $-2°$ and 4°C.

Figure 6.10 has examples of coarse-scale T–S–V diagrams. Fine-scale diagrams were also prepared with grid squares of 0.5C° and 0.1% side to show more detail (see the Montgomery, Cochran and Pollak references) and, more recently, Worthington (1981) has presented T—S—V diagrams with 0.1C° and 0.01% squares for the deep waters. Some of his results are shown below.

Similar diagrams were prepared to show the T–S–V characteristics of the world ocean (Fig. 6.10(b)). Montgomery pointed out that the mean values calculated from the 1958 statistics did not differ significantly from the values determined by Krümmel in 1907. One of the chief reasons for the small differences between Krummel's averages based on few data and Montgomery's based on many more is the small range of temperature and salinity found in the bulk of the oceans. The oceanographic "climate" is so uniform below the surface layer that comparatively few observations sufficed to enable Krümmel to make a good estimate of the mean values. It must not be concluded, however, that little had been gained in oceanographic knowledge between 1907 and 1957. Mean values conceal much detail, and the very fact that deep-water characteristics have a small range of values makes it difficult to distinguish one water mass from another. In fact, the development in recent years of the electrical conductivity salinometer, with its increased sensitivity and precision, has helped a great deal to distinguish between different water masses, even though the increased sensitivity may not result in much change in mean values. Worthington's 1981 census of the world ocean water masses was based on such high-quality salinity data, which is presumed to be the reason why some of his volume figures differ somewhat from those of Montgomery. Worthington pointed out that while the North Atlantic had plenty of high-quality stations, they were sadly lacking in a large part of the Pacific and in the South Atlantic and South Indian Oceans.

The usual way to present the T–S–V diagram on paper is to display temperature and salinity along two axes and volume as numbers, as in the examples above. As the diagram presents relations between three quantities we would like to present it in three dimensions. It would not be difficult to build up a three-dimensional T–S–V model by placing on each square a column

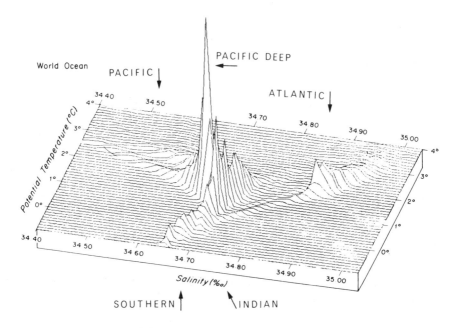

FIG: 6.11.   *Simulated three-dimensional T–S–V diagram for the cold water masses of the World Ocean.*

whose height represented volume, although such a model would obviously only be suitable for demonstration purposes. An alternative nowadays is to show a computer-generated simulation view of such a three-dimension model. Figure 6.11 is an example of such a simulation for the cold waters of the oceans from the new census of world ocean waters by Worthington (1981) for classes of $0.1K \times 0.01\%$. The ridge extending to the lowest temperatures at about $34.6\%_{oo}$ represents water in the Southern Ocean, the ridge extending to high temperatures and salinities is for the Atlantic, while the Pacific Ocean water is represented by the ridge extending toward $3°C$ and $34.4\%_{oo}$. The central spike, representing about 2.6% of the cold water is for the deep water in the North Pacific. Indian Ocean waters are represented in the central ridge from about $-0.5°$ to $2°C$.

While Fig. 6.11 shows the volume distribution of the cold (deep) waters, Fig. 6.12 shows the distribution of water properties for the upper, warm ($> 4°C$) water which constitutes about 24% of the total ocean volume. The contours show the volumes (in 10^3km^3) in classes of $0.1K \times 0.01\%$. The shaded area at the bottom represents the remaining 74%, the cold water

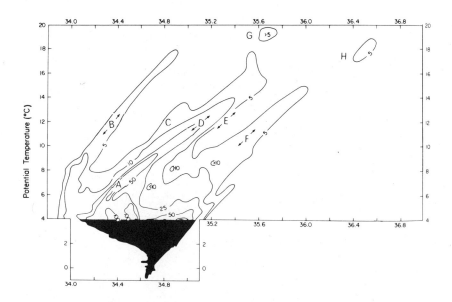

FIG. 6.12. *Contoured T–S–V diagram for the warm water masses of the World Ocean: (A) Subantarctic, (B) North Pacific and Eastern South Pacific, (C) South Pacific and Subtropical North Pacific, (D) South Atlantic, South Pacific and Indian, (E) South Atlantic and Indian, (F) Western North Atlantic, Red Sea Water in Indian Ocean, (G) South Pacific, (H) 18° Water in Western North Atlantic. (The black shaded area at the bottom represents the cold water of Fig. 6.11.)*

shown in Fig. 6.11. The ocean areas where much of the warm water occurs are indicated on the three prongs of the contours in the figure.

Before leaving this description of the T–S–V diagrams, the reader should note that the temperature used, although indicated by T in the abbreviation, is usually the potential temperature θ as the water volumes usually come from a wide range of depths.

The last form of three-dimensional characteristic diagram considered is the temperature–salinity–time (T–S–t) diagram which is a compact way of showing the sequence of combinations of water properties with time. As an example, Fig. 6.13 shows monthly mean values for three zones of the Australian Great Barrier Reef lagoon (Pickard, 1977). In the south, the annual variation is mainly in temperature, in the north there are large variations of both temperature and salinity, while in the centre zone there is an extreme salinity variation. The reason for the differences is that the north and centre zones are subject to heavy monsoonal rains in the austral summer (January to April) while the south zone escapes these. The very low salinity in the centre is due to the rivers there which drain much larger inland areas than do the smaller

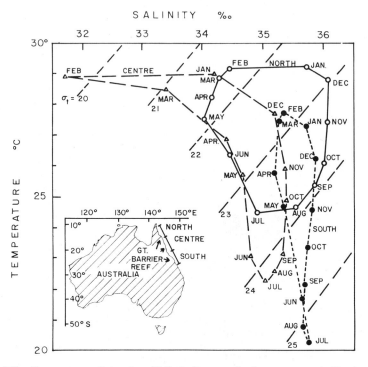

FIG. 6.13. *Temperature–salinity–time (T–S–t) diagrams for lagoon waters inside the Great Barrier Reef.*

rivers in the north. The temperatures used for these surface T–S–t diagrams are the *in situ* values T.

6.35 Conclusion

From the above descriptions, it will be seen that it is not practicable to recommend any one form of graphical presentation as the standard for oceanographic data. An oceanographer who wishes to present his data and his analysis of it in a report has to decide for himself which form best displays the features which he considers to be important. He may well have to use all of the methods mentioned to make all his points, or he may have to invent new ones.

CHAPTER SEVEN

Circulation and Water Masses of the Oceans

WE NOW come to the circulation of the oceans, first discussing some of the general features of the main ocean circulations and then proceeding to describe the circulation and character of the water masses in the individual oceans.

7.1 Introduction

It must be stressed that in one way or another the energy from the sun's radiation is responsible for these circulations. In fact, both the atmospheric and the oceanic circulations are driven by this energy and we should perhaps really study both together (e.g. Perry and Walker, 1977). As our book is primarily about the ocean we will concentrate on the water circulation but it will be apparent that the atmosphere plays a considerable part in driving ocean currents.

The ocean circulation can be divided into two parts, the thermohaline and the wind-driven components. Another way to describe these is to say that the ocean circulation is in part due to changes in density caused by weather or climatic changes and in part due to the wind stress. The manner in which these factors determine the circulation is discussed in texts on dynamical oceanography and all that will be done here is to describe the physical processes in a qualitative fashion.

7.11 Thermohaline circulation

The term *thermohaline circulation* is used to refer to the movement of water that takes place when its density is changed by a change of temperature or of salinity in a suitable part of its bulk. A standard laboratory demonstration of thermal circulation is to take a large beaker of water and heat it at the bottom by means of a bunsen burner. The water which is heated expands and rises (convection) in accordance with Archimedes' Principle. A few crystals of dye dropped into the water will dissolve and colour the bottom water and show it rising over the source of heat and sinking elsewhere. The sinking in this case is simply a consequence of continuity of volume. In the case of the atmosphere the sun's energy is absorbed by the land, this heats the air near to it and a similar circulation to that in the beaker takes place. Cumulus clouds developing on a summer day often reveal where the upward flow of this kind of circulation is taking place.

The laboratory experiment with the beaker of water and the heating of the atmosphere have this in common—the heat is supplied at the bottom of the fluid. The situation is different in the ocean in that the sun's heat is supplied at the top; this makes a great deal of difference. A liquid warmed at its top surface cannot rise and therefore a circulation does not develop. If we were to take a long tank of water and heat the surface at one end, then some motion would take place. The heated water would expand and raise the water level slightly at this end. The warmer surface water would then flow toward the cold end over the top of the cold water but no circulation in the full sense would develop. Some of the early theories of the ocean circulation ascribed the currents to a thermal flow of this sort. The stronger heating of the sea at low than at high latitudes was recognized and it was suggested that the water then flowed north and south toward the poles. While it is possible that the differential heating between low and high latitudes may have some effect of this nature it is now believed to be a minor one.

Since there is no significant source of heat at the bottom of the oceans a thermal circulation like that in the beaker and in the atmosphere plays little part in the ocean circulation.

However, heating at low latitudes is only one part of the heat budget. It has already been shown that there is a net heat loss from the oceans at high latitudes. The result of the cooling of the water is an increase in density which may be sufficient to cause it sink and so displace the deeper water. The laboratory experiment to illustrate this is to float an ice cube in a beaker of water; the water cooled by the ice will sink. (Allow the water to sit in the beaker for a few hours to allow any motion to die out before adding the ice cube, and put a drop of dye on the latter to reveal the cooled water sinking to the bottom.) If a little salt has been dropped to the bottom of the water some hours beforehand, it will have dissolved and increased the density of the bottom layer. Then the dyed water will stop sinking at the top of this "bottom water". The thermohaline circulation of the oceans is due to an increase of density at the upper surface, either directly by cooling or indirectly when ice freezes out, ejecting salt and thus increasing the density of the remaining water. In the North Atlantic the cooling effect in the winter is considered to be responsible for the sinking of water to considerable depths. In the Antarctic, the freezing effect is important. The sea-ice will not be pure ice, as some salt is usually trapped among the ice crystals, but it is usually less saline than the sea-water from which it was formed. The remaining sea-water is thus more saline and more dense than before, and this will cause it to sink. One may summarize then by stating that a characteristic of the thermohaline circulation of the oceans is that it originates as a vertical flow sinking to mid-depth or even to the ocean bottom, followed by horizontal flow.

An increase of salinity may also occur due to evaporation in the tropics but generally this is accompanied by solar heating and the decrease in density due to heating prevails over the increase due to evaporation. The more saline water stays near the surface. Evaporation is therefore not a primary factor in causing water to sink but it does act indirectly as will be described for the Mediterranean and the Red Sea.

7.12 Wind-driven circulation

The *wind-driven circulation* is principally in the upper few hundreds of metres and therefore is primarily a horizontal circulation in contrast to the thermohaline one. In texts on dynamical oceanography the mechanism is described by which the wind blowing over the sea surface causes it to move. Here we will just accept this as a consequence of fluid friction, but note that the resultant direction of motion in the open ocean is not the same as that of the wind. As will be described in an elementary fashion in Section 8.2, the rotation of the earth gives rise to the "Coriolis force" which causes the wind-driven currents in the upper, mixed layer in the open sea to move in a direction to the right of the wind direction in the northern hemisphere and to the left in the southern.

Ocean currents are then a result of the combined effects of the thermohaline motions and of the wind-driven ones. The former probably prevail in the deep water while the latter prevail in the upper waters. In both cases, the motion usually continues far beyond the place where it was initiated, just as water may be made to circulate round a hand basin by blowing along the surface at one side only. Also it should be appreciated that while the wind primarily causes horizontal motion, there may be consequent vertical motion such as the up- and down-welling in the equatorial zone (Section 7.613) and upwelling near the coast (Section 8.2). Similarly, while the thermohaline effects primarily give rise to motion with a vertical component, horizontal motions may result from continuity requiring inflow or outflow, and from changes in the field of mass (i.e. density) leading to geostrophic currents. Theoretical studies have demonstrated that thermohaline forces alone can give rise to major anti-cyclonic (clockwise in the northern hemisphere) gyres which are similar to, but weaker than, those generated by the wind alone. For further discussions of the roles of the two circulations see Stommel (*The Gulf Stream*, 1965) and Wyrtki (1961).

7.13 Circulation and water masses

When one attempts to describe the *circulation* and *water masses* of the oceans one is faced with a problem. If one wishes to present the details one can only do so for a region of limited extent and so one must divide the world ocean into regions for this purpose. In practice, this is not too difficult to do; the oceans almost divide themselves into regions. For instance, the Atlantic can be described regionally as the North, the Equatorial and the South Atlantic, together with the adjacent seas such as the Mediterranean, the Labrador Sea, etc. A professional oceanographer who is studying a particular region for a specific purpose, e.g. as part of a study of the fisheries, will be concerned with the details of that region. But almost invariably when he selects a particular area for study he finds that the waters within the region are influenced by waters without the region, e.g. by currents entering the area of study from elsewhere. Then to understand the particular area he has to extend the study beyond that area, and sometimes it is difficult to stop this extension before reaching the opposite boundary of the ocean. In other words, to understand the part one must understand the whole.

Since the present book is intended only as an introduction to Descriptive Oceanography, and since the total volume of detailed data available is far too great to include or even summarize within its pages, the main endeavour in the following description will be to acquaint the reader with the major features of the ocean circulations and water masses, with some description of details for a few areas as examples. It is hoped that this approach will provide him with enough feeling for the character of the ocean circulation and water masses as a whole that he will have sufficient background to study smaller areas in detail later on, e.g. using Tchernia's text *Descriptive Regional Oceanography*.

Again, when describing the circulations and the water masses one is faced with the question of which to present first. It is the old problem of "which came first, the chicken or the egg?" In this book the authors have decided to take an operational approach. For the upper waters, the circulation will be described first but for the deep waters the water masses will be described first. The reason for this procedure is that we have a good deal of direct information about the circulation of the upper layers, but the movements of the deep waters have largely been inferred from the distributions of properties.

From the analysis of ship's navigation logs, as described in Chapter 6, we have a knowledge of the surface circulation of much of the oceans. This knowledge is substantial and detailed in the regions much travelled by ships, such as the main traffic routes across the North Atlantic and North Pacific, but is scanty in other regions such as the eastern South Pacific and the southern Indian Ocean. As has already been stated, the upper-layer circulation is driven primarily by the winds and from the results of studies in dynamical oceanography we can obtain from the surface-layer circulation a very good idea of the whole upper-layer circulation down to the thermocline. Therefore, in describing the upper layers of the ocean, the circulation will be described first and the water properties second as these are chiefly determined by the history of the water, i.e. where it has been carried by the currents.

For the deep water, the water mass characteristics will be described first and then the movements, since most of our knowledge of the deep-water circulation has been obtained by interpretation of the distributions of properties. In addition to the deep water flows having been inferred from the property distributions the flows are mostly driven by the distribution of density which is itself determined by the distributions of temperature and salinity. Only in recent years has the invention of the neutrally buoyant float and the development of mooring techniques and current meters permitted satisfactory direct observations of the deep water movements. To date, the information available from these sources is limited in amount and is from only a few regions.

Some indication of what is meant by "upper" and "deep" waters must be given before going further. It is not easy to give an exact figure for the thickness of the *upper water* but it is usually taken to extend from the sea surface to the depth at which the decrease in temperature with depth becomes small. This may be between 300 and 1000 m (Fig. 4.5). The upper layer contains the surface or *mixed layer* of 50- to 200-m depth in which most of the seasonal variations of properties occur and which is usually fairly homogeneous due to mixing by the action of the waves caused by the wind. The remainder of the upper layer is usually well stratified and stable, and includes the thermocline. The *deep water* includes all that below the upper water, and is less stable than the latter. If the layer in contact with the sea bottom has properties distinct from those of the deep water above, it is referred to as the *bottom water*.

Before proceeding to the more detailed description of the ocean areas, the main features will be presented briefly so that the reader may appreciate how the details fit into the whole. The *circulations in the major ocean areas* show considerable similarities; the differences are largely in detail. In the upper layer there is a major clockwise circulation or "gyre" in both the North Atlantic and the North Pacific and a counter-clockwise gyre in the southern parts of the Atlantic, Pacific and Indian Oceans. These "anticyclonic" circulations dominate the low- and mid-latitude portions of these oceans. In the North Atlantic and Pacific a very conspicuous feature is that the currents are narrower and swifter on the west side of each ocean than elsewhere (*westward intensification* of the currents). There is evidence of the same phenomenon in the South Atlantic and Indian Oceans but the western South Pacific circulation is rather complex and the intensification is not clear. In the equatorial regions of all three oceans there are similar current systems consisting of a westward-flowing South Equatorial Current at or south of the equator and a westward-flowing North Equatorial Current further north. In the Pacific these two are separated by an eastward-flowing North Equatorial Counter Current along the full width of the ocean; in the Atlantic the Counter Current is only significant in the eastern part. These equatorial counter-currents owe their existence to variations in the wind stress from north to south across the equator. In the Indian Ocean, the three-current pattern is present for part of the year and a two-current system for the remainder. Another component of the equatorial current system is the subsurface Equatorial Undercurrent which flows eastward along the equator, at about 100 m depth, carrying water built up in the west by the westward-flowing Equatorial Currents. Defant called the surface equatorial current systems the "backbone of the circulation" to emphasize our belief that the northern and southern gyres are driven mainly by the trade winds in low latitudes. The facts that the regions have maintained their identity of circulation and of water mass characteristics for more than a hundred years indicates that the climatic processes which determine them must be continuing to act.

In the deep water the major flows are north and south, not necessarily evenly distributed across the widths of the oceans but probably stronger on the west sides.

The two polar regions show marked differences which are due in part to the differences in character of the driving forces but principally to the difference in topography of the two basins.

Since the Southern Ocean is openly connected with all the other oceans, and water which acquires its characteristics here has a profound influence on the deep waters in the other oceans, we will start with this region.

7.2 Southern Ocean

7.21 Divisions of the Southern Ocean

The Southern Ocean has the land mass of the Antarctic continent to form its southern boundary but has no land boundary to its north and is continuous with the other oceans. However, the surface waters of the region have well-defined characteristics whose isopleths run roughly parallel to lines of latitude (i.e. are zonal) and features of these characteristics are used to define the northern boundary of the Southern Ocean. The classical picture of the upper layer zones is as follows. Going north from the Antarctic continent the average sea surface temperature increases slowly until a region is reached where a relatively rapid increase of 2 to 3 K occurs. The surface water from south of this region is moving north and sinks when it reaches the region, continuing north below the surface. At the surface therefore the water is converging to this region which was known as the Antarctic Convergence but is now called the *Antarctic Polar Front* (APF) (Fig. 7.1). Continuing north

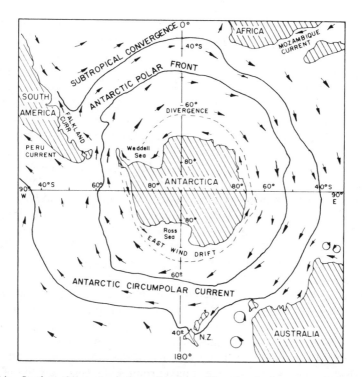

FIG. 7.1. *Southern Ocean circulation and mean positions for the Antarctic Polar Front and Subtropical Convergence.*

from this APF the temperature rises slowly to a second region where it rises rapidly by about 4K and the salinity by about 0.5‰. This is referred to as the *Subtropical Convergence* (Fig. 7.1).

The APF is found at about 50°S in the Atlantic and Indian Oceans and at about 60°S in the Pacific (Fig. 7.1). The Subtropical Convergence is at about 40°S round most of the Antarctic but its position has not been well determined in the eastern South Pacific which is a poorly known region oceanographically. The two convergences divide the surface waters of the Southern Ocean into two zones, the *Antarctic zone* from the continent to the APF and the *Subantarctic zone* from there to the Subtropical Convergence (Fig.7.1). In the Antarctic zone, the surface temperature is between −1.9° and 1°C in winter and between −1° and 4°C in summer, while in the Subantarctic zone it is between 4° and 10°C in winter and up to 14°C in summer. (Note that unless a minus sign is placed before a temperature it is understood to be positive.)

The above description of the upper layer of the Southern Ocean was based on data available to the mid-1940s. Subsequently more data have become available, both from oceanographic expeditions and from supply ships to the Antarctic continent, particularly in the sector from western Australia east to South America. It now appears that the characteristics, particularly in the 50–60°S zone, are more complicated than described above. In the first place a single, clearly defined convergence of surface waters at the boundary between the Antarctic Surface Water and the Subantarctic Surface or Upper Water (both described later) (Figs. 7.1 and 7.2) is not confirmed; rather the transition zone between these water masses extends over a range of latitudes with seasonal and longitudinal changes in width and position. The term *Antarctic Polar Frontal Zone* (APFZ) is preferred to "Antarctic Convergence" and from studies of both station data (temperature and salinity) and XBT sections (temperature only) Emery (1977) suggested the terms *Antarctic Front* for the southern boundary (earlier the APF) and *Subantarctic Front* for the northern boundary of this zone. The northern front limits the southward extension of Subantarctic Surface Water and is marked by a strong temperature gradient as well as a shift in T–S values. A change in T–S characteristics also occurs at the Antarctic Front which, as the northern limit of Antarctic Surface Water, coincides with a subsurface temperature minimum in summer. Where there is a northern land boundary, i.e. Australia, New Zealand and South America, these fronts are often associated with zonal current jets. The structure of the zone itself is quite variable and complex, with evidence of eddies and meanders. Much of this variability, and indeed the width of the Frontal Zone itself, is due to instabilities in the zonal current jet associated with the APF. This front has been observed to meander, forming northward extensions which entrain and capture Antarctic Surface Waters. These meanders then separate to form cold, cyclonic eddies which populate the APFZ.

7.22 Southern Ocean circulation

The unique geography of the Southern Ocean makes it the only place where ocean currents run completely around the globe. In a narrow zone round most of the continent there is a westward-flowing coastal current, called by Deacon (1937) the *East Wind Drift* because it is attributed to the prevailing easterly winds near the coast. Farther north the remainder of the Southern Ocean is dominated by a strong, deep eastward-flowing current known as the *Antarctic Circumpolar Current* (ACC) (Fig. 7.1).

The ACC has been referred to as the "West Wind Drift" which, as the flow is to the east, may require explanation. The reason arises from the different manners in which wind directions and water current directions are stated. It is usual to express wind direction as where the wind comes *from*, i.e. the westerly winds blow from west to east. This convention results from sailing-ship experience in which one is more conscious of where the wind and weather are coming from than where they are going to. For water currents, however, it is usual to state the direction *toward* which the water is flowing, e.g. the eastward Circumpolar Current flows to the east. (The westerly wind in the Southern Ocean was notorious in sailing ship days and, together with the eastward current, made it difficult for such vessels to round Cape Horn from the Atlantic to the Pacific.)

The surface flow of the ACC is driven primarily by the frictional stress of the westerly wind which led to the earlier name of West Wind Drift. The wind stress, combined with the Coriolis force, also contributes a northward component to the surface current resulting in the formation of fronts, i.e. convergences, within the APFZ. Below the wind-driven surface layer, the density structure appears to be in geostrophic balance with the circulation.

In its circuit round the continent the ACC is obstructed only in the narrow Drake Passage between South America and the Palmer Peninsular projecting north from Antarctica. In general, the ACC is not very fast with surface speeds of only about 4 cm/s in the Antarctic zone increasing to 15 cm/s north of the Polar Front and then decreasing again toward the Subtropical Convergence. However, the current is very deep and its volume transport was estimated by Sverdrup as up to 110 Sv (1 Sv = 1 sverdrup = 10^6 m^3/s which is the commonly used unit for expressing the volume transport of ocean currents. The "sverdrup" is not an SI unit but will be used for brevity.) Other workers have determined transport values, at various times, from small westward flows to as much as 237 Sv eastward (the usual direction). The earlier estimates were from geostrophic calculations made from the density distribution obtained from measurements of water temperature and salinity made during periods of only a few days at a time and with the usual uncertainties in converting the relative velocity profiles to absolute values. Some had limited current measurements to assist this conversion but most relied on "depth of no motion" assumptions. Nowlin *et al.* (1977) used both geostrophic and current meter measurements and obtained transports during a period of 3 weeks ranging from 110 to 138 Sv and averaging 124 Sv. From year-long current meter records, Bryden and Pillsbury (1977) obtained an average value of 139 Sv but during the year the transport varied from 28 to 290 Sv. Typical time and space scales for variations were 2 weeks and less than 80 km (compared with the 800 km width of the Drake Passage). Many of these variations are associated with the strong zonal jets at the major fronts. In these jets, current speeds of 50 to 100 cm/s have been observed.

Most of the transport measurements have been made in the Drake Passage because here the current is clearly limited north and south and the width is at its minimum value. As part of the International Southern Ocean Studies (ISOS) a dense array of current meters was deployed for a year, from which it is hoped to obtain an adequate description of the short- and long-term variations of the ACC which, at its maximum, is clearly the mightiest current in the oceans. It shows some variations in direction as it flows round the continent and there is evidence that some of these are due to the effects of the submarine topography. Some of the current branches off and flows north between Australia and New Zealand at subsurface levels and some, including surface water, flows north as the Peru Current up the west coast of South America into the South Pacific (Fig. 7.1), making significant contributions to the circulation and water masses of that ocean. There is some flow from the ACC northward into the Atlantic between South America and the Falkland Islands as the Falkland Current (Fig. 7.1).

7.23 Southern Ocean water masses

7.231 Antarctic zone

The water masses of the Southern Ocean (Fig.7.2) have typical high latitude characteristics. South of the APF the *Antarctic Surface Water* has properties which are determined by ice melting in summer and by cooling in winter. This layer of 100- to 250-m thickness has a salinity of less than 34.5°/$_{oo}$ and a low temperature ranging down to the freezing point, about $-1.9°$C at that salinity. The temperatures are lowest in the south and increase toward the north due to the absorption of heat during the summer. It must be

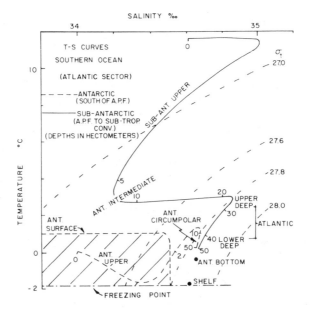

FIG. 7.2.   *T–S curves for Southern Ocean waters, Atlantic sector.*

remembered that in the polar regions, south or north, an effect of sea-ice is to limit the range of temperature variation between winter and summer. In one direction, the freezing point of sea-water at the salinities found at high latitudes is no lower than $-1.9°C$ and limits the reduction of temperature which can occur in winter. In the other direction, the melting of ice requires a considerable proportion of the heat inflow during the summer, because of the large latent heat of melting of ice, leaving only a small part to raise the temperature of the water.

Below the surface and extending to the bottom at depths to 4000 m is the *Antarctic Circumpolar Water*. Its temperature in the Antarctic zone rises to a maximum of 1.5 to 2.5°C at 300 to 600 m and then decreases to between 0 and 0.5°C at the bottom. Its mean salinity is close to $34.7°/_{oo}$. These characteristic properties are found all round the Antarctic continent at the same depths, evidence that this water is carried around the continent by the ACC.

A very important water mass is the *Antarctic Bottom Water*. Most of it is formed in the Weddell and Ross Seas in the Antarctic continent (Gill, 1973) (Fig. 7.1). Its water is a mixture of the Antarctic Circumpolar Water with *Shelf Water* which has attained its properties on the continental shelf region of these seas. The Shelf Water has a temperature of $-2.0°C$ (the freezing point decreases with increase in pressure) and a salinity of 34.4 to $34.8°/_{oo}$. Its density (σ_t) of 27.96 is among the highest found in the southern oceans. (Values to 28.1 are found in the Ross Sea.) The formation of these dense waters has been the subject of much discussion; a review is presented by Warren (1981). The mixing of Shelf Water with Circumpolar Water reduces its density to slightly below 27.9 as it flows out from the Weddell Sea. Because of this high density it flows down the continental slope into the South Atlantic (Fig. 7.3) and also eastward through the Indian and Pacific sectors of the Southern Ocean. The eastward flow round the continent is deduced from the continuity of temperature and salinity values together with the steady decrease in oxygen content from the Atlantic sector through the Indian and Pacific sectors. If we knew the rate at which oxygen is used up in the water it would be possible to estimate the speed of flow from the rate of decrease of oxygen. Unfortunately we do not have any reliable figures for the rate of oxygen consumption in this region.

7.232 Subantarctic zone

In the Subantarctic zone the water masses become more numerous (Fig. 7.2, full line). Some of them originate in the Southern Ocean but others originate outside and information which will be given later in this chapter has been used to identify them. In addition, the upper water masses show some differences between the Atlantic, Indian and Pacific sectors, in contrast to the relatively uniform character of the Antarctic surface water.

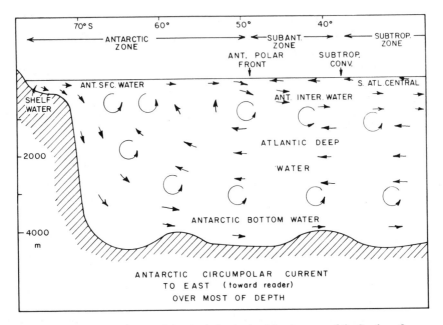

FIG. 7.3. *Vertical components of the circulation in the Atlantic sector of the Southern Ocean.*

The *Subantarctic Upper Water* occupies the upper 500 m or so and has a temperature of 4 to 10°C in the winter and up to 14°C in summer, and a salinity from 33.9 to 34.9°/$_{oo}$ in winter and as low as 33°/$_{oo}$ in summer as ice melts. The lowest temperatures and salinities are found in the Pacific sector and the highest in the Atlantic sector. This Upper Water has a southward component of motion which accounts for its higher temperature than the waters farther south. A salinity maximum is also present in all sectors at 150 to 450 m depth.

Below this water is the *Antarctic Intermediate Water* which includes surface water from the Antarctic zone and is formed by mixing below the surface in the APFZ. It continues northward with some admixture from the Upper Water (Fig. 7.3). The Intermediate Water has a thickness of about 500 m at the northern limit of the Subantarctic zone, under the Subtropical Convergence. It is more homogeneous in properties than the Upper Water, with a temperature of 2 to 3°C and a salinity of 34.2°/$_{oo}$. It sinks because its density (σ_t) of 27.4 is greater than that of the water to its north. As it flows to the north it mixes with more saline water above and below and its salinity rises gradually. The Antarctic Intermediate Water flowing north forms a tongue of relatively low salinity water with its core at 800 to 1000 m depth at 40°S and to the north of this (see Fig. 7.9). The water also has a relatively high oxygen content of 5 to 7mL/L since it has only recently left the surface.

The next water mass below is the *Deep Water* which moves with a southward component of motion in addition to its eastward circulation as a part of the ACC (Fig. 7.3). It lies between 1500 and 3000 m depth in the Subantarctic zone and has a temperature of 2 to 3°C and a salinity of 34.7 to 34.9°/$_{oo}$ and an oxygen content of 4 to 5 mL/L in the Atlantic sector. It is sometimes referred to as the "warm" deep water. In a vertical profile and in the T–S diagram (Figs. 7.2, 7.14(a)) the water does show a very slight temperature maximum but it is only warm relative to the colder Intermediate and Bottom Waters. It was first observed in 1821 but only recognized as being of North Atlantic origin by Merz and Wüst (1922). The Upper Deep Water shows a salinity maximum in the Atlantic and Indian Ocean sectors (Fig. 7.14) while the Lower Deep Water is difficult to distinguish from the Antarctic Circumpolar Water. In the Pacific sector (Fig. 7.14) the Upper Deep Water is less easy to distinguish from the Antarctic Circumpolar Water than in the other sectors. The Deep Water has a southward component of motion, contrary to the Intermediate Water above and the Bottom Water below. There is inevitably some mixing between them, and the waters above and below are both modified by mixing with the Deep Water. In moving south from the Atlantic, the Deep Water rises toward the surface (Fig. 7.3). It is probable that it does not actually reach the surface as a water mass but diverges north and south, mixing with surface water and contributing to the Antarctic Intermediate Water and to the Bottom Water. The Deep Water comes from the North Atlantic and has been away from the surface for a long time; it has a relatively low oxygen content by Atlantic standards and high concentrations of nutrients. This continual supply of nutrients is one of the reasons for this region being prolific in phytoplankton (plant) growth and consequently in zooplankton. The latter is a source of food for larger animals in the sea and the whaling industry of the Southern Ocean was one consequence. In a vertical section of water properties (e.g. Fig.7.9) the Deep Water is characterized by an oxygen minimum as it lies between the Intermediate and the Bottom Waters, both of which were more recently at the surface and consequently are richer in oxygen.

The deepest water in the Subantarctic zone is the Bottom Water flowing north (Fig. 7.3). Mixing with the overlying Deep Water its temperature is raised to 0.3°C and its salinity to 34.7 to 34.8°/$_{oo}$.

Summarizing, it is possible to divide the Southern Ocean into three zones round the continent, an Antarctic zone nearest to the land and a Subantarctic zone farther north, separated by a transitional region called the Antarctic Polar Frontal Zone. In the Antarctic zone (Fig. 7.1) there is a thin surface layer of cold Antarctic Surface Water of relatively low salinity from summer melting of ice. Below this there is the large mass of the Antarctic Circumpolar Water of uniform properties all round the continent. The cold Antarctic Bottom Water formed at the continental edge flows north and east into the three main oceans. In the Subantarctic zone there are four distinct water masses. The Subantarctic Upper Water is warmer and more saline than the Antarctic Surface Water. Below this the Antarctic Intermediate Water is fed by lower salinity Antarctic Surface Water flowing north and descending beneath the Upper Water. Below this, marked by a salinity maximum, is the large body of Deep Water flowing south and rising over the Bottom Water flowing north. The Intermediate and Bottom Waters are relatively high in oxygen content from their recent contact with the surface, or with surface water, while the Deep Water is relatively low in oxygen content. The Deep Water can be divided into an Upper Deep of slightly higher temperature and salinity than the Lower Deep Water.

7.3 Atlantic Ocean

7.31 Atlantic Ocean as a whole

The *upper water circulation* of the Atlantic Ocean as a whole consists in its gross features of two great anticyclonic circulations or "gyres", i.e. a counter-clockwise one in the South Atlantic and a clockwise one in the North Atlantic (Fig. 7.4). At first sight it might seem appropriate to liken these two gyres to two gear wheels revolving and meshing near the equator. This, however, gives the impression that one may be driving the other, which is not the case. Rather the two gyres are driven separately, each by the trade winds in its own hemisphere (similar though not identical to those in the Pacific, Fig. 7.24), and they are separated over part of the equatorial zone by the eastward flowing Counter Current.

7.32 South Atlantic Ocean

7.321 South Atlantic circulation

In the South Atlantic the upper-water gyre extends from the surface to a depth of about 200 m near the equator and to about 800 m at the southern limits of the gyre at the Subtropical Convergence. The different portions of this gyre have different water properties and have individual names which are given in Fig. 7.4. It is considered that the wind stress of the south-east trade winds between the equator and 10 to 15°S is the main driving force. This acts upon the sea and causes the *South Equatorial Current* to flow west toward the American side of the South Atlantic. Part of the current passes across the equator into the North Atlantic and will be discussed later. The remainder turns south along the South American continent as the *Brazil Current* which then turns east and continues across the Atlantic as part of the Antarctic Circumpolar Current and then turns north up the African side as the *Benguela*

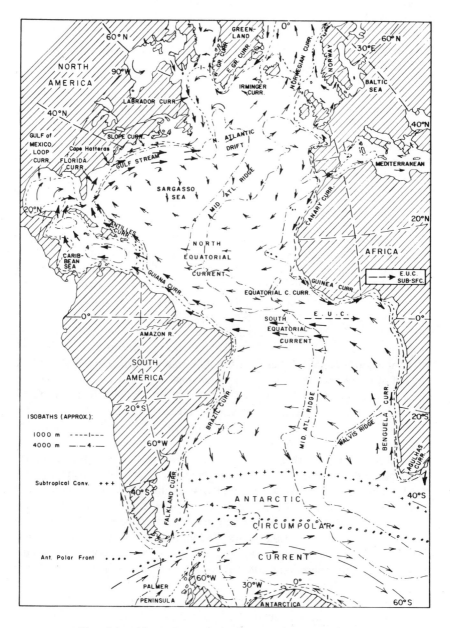

FIG. 7.4. *Atlantic Ocean—bathymetry and surface circulation.*

Current. The Brazil Current is warm and saline, having come from the tropic region, while the Benguela Current is cold and less saline because of the contribution of Subantarctic Water and of upwelling along the African coast. A contribution to the water in the South Atlantic comes from the *Falkland Current* flowing north from Drake Passage up the coast of South America and separating the Brazil Current from this coast to about 30°S. The South Atlantic circulation is bounded on the south by the Subtropical Convergence. For simplicity this Convergence and the Antarctic Polar Frontal Zone are shown in Figs. 7.1 and 7.4 as lines but, as described for the Polar Frontal Zone, they must be considered in reality to be zones of finite extent with seasonal variations of position.

7.322 South Atlantic volume transports

Wüst (1957) made calculations of the volume transport of water in the South Atlantic and these are interesting both as descriptive of that ocean and as examples of the application of oceanographic principles. When considering Wüst's calculations it must be remembered that there were no direct current measurements of any consequence available to him, and the volume transports had to be deduced indirectly. Calculation by the geostrophic method from the density distribution gave the relative currents from top to near bottom across the ocean. To make the relative currents absolute a depth of no motion was chosen between the Antarctic Intermediate Water which must flow north and the Deep Water which must flow south. This depth varies over the ocean but is at about 1400 m for a section at 30°S from the South American to the African shore. These absolute currents could then be used to determine the flow at various depths and positions across the section. The resulting volume transports of the different water masses had to satisfy volume continuity, i.e. there must be no net flow north or south through the trans-oceanic section because there is no evidence of any long-term change of sea-level north or south of the section. It also assumes that there is no net flow into or out of the North Atlantic + Arctic Sea combination. (There are certainly large exchanges of water between the Atlantic and Arctic but these are internal to the present discussion and cannot affect the mean sea-level.) The assumption of no net flow into the Atlantic + Arctic combination is not quite true because there is some flow through the Bering Strait from the Pacific into the Arctic. Recent measurements and estimates suggest that the mean transport ranges from 1 to 2 Sv, but this is small compared with the volume transports across the section in the South Atlantic and can be ignored. It should be noted that conservation of salt across the section also had to be satisfied.

Sverdrup summarized the volume transports (*The Oceans*, 1946) and these are shown schematically in Fig. 7.5. In the upper layer there is a transport to the north of about 23 Sv on the east side (Benguela Current and South Atlantic Gyre) as against 17 Sv to the south on the west side (Brazil Current and the Gyre). The Antarctic Intermediate Water has a transport to the north of 9 Sv, the Deep Water has a transport to the south of 18 Sv, and the Bottom Water has a transport to the north of 3 Sv. There is therefore a net transport to the north of 18 Sv in the upper layers (from the surface to 1400 m depth) together with the Bottom Water, balanced by 18 Sv to the south in the Deep Water. This represents a fairly rapid rate of exchange of water in the north–south direction in the South Atlantic Ocean.

An example of a recent calculation of heat and fresh-water fluxes was given in Section 5.373. In that case continuity of fresh water relative to a standard salinity was used instead of conservation of salt relative to zero salinity.

7.33 North Atlantic circulation: general

The North Atlantic has been subjected to intensive study, particularly during the last 20 years and, as a consequence, some new ideas about its circulation have developed recently. The classical pattern will be presented first and then new interpretations of the data will be described with some criticisms.

The circulation of the North Atlantic shown in Fig. 7.4 follows that presented by Sverdrup *et al*, in *The Oceans* (1946) and by other authors earlier (e.g. Iselin, 1936). The clockwise gyre (Fig. 7.4) may be considered to start with the *North Equatorial Current* driven by the north-east trade winds. This

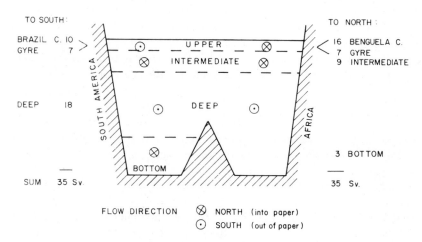

FIG. 7.5.   *South Atlantic—north–south volume transports (schematic).*

current flows to the west and is joined from the south by that part of the South Equatorial Current which has turned across the equator into the North Atlantic. Part of this combined flow goes north-west as the *Antilles Current* east of the West Indies, and part goes between these islands and through the Caribbean and the Yucatan Channel into the Gulf of Mexico. In its passage through the Caribbean the flow is driven by the east winds in this region and the water piles up in the Gulf of Mexico. From there it escapes between Florida and Cuba into the North Atlantic as the *Florida Current.* The characteristics of this current indicate that its source is chiefly the North and South Equatorial Current waters which traverse the Caribbean. Little of the water from the Gulf of Mexico itself appears to be carried out in the Florida Current. Studies have shown, however, that the flow through the Gulf of Mexico sometimes forms a large loop. This loop often spawns an anticyclonic eddy which moves westward in the Gulf. Off the coast of Florida the Current is joined by the Antilles Current and from about Cape Hatteras, where the combination breaks away from the North American shore, it is called the *Gulf Stream.* The Florida Current water is distinguished by a salinity minimum due to the Antarctic Intermediate Water (Fig. 7.9) brought in by the South Equatorial Current. The Antilles Current is composed mostly of North Atlantic Water and the salinity minimum is much less evident in it. The Gulf Stream flows north-east to the Grand Banks of Newfoundland at about $40°$ N, $50°$ W. From there, the flow which continues east and north is called the *North Atlantic Current.* This divides and part turns north-east between Scotland and Iceland and contributes to the circulation of the Norwegian, Greenland and Arctic Seas which will be described later. The remainder of the North Atlantic Current turns south past Spain and North Africa to complete the North Atlantic Gyre and to feed into the North Equatorial Current. The southward flow must not be thought of as a narrow stream along the European side. In fact it covers the greater part of the North Atlantic as is shown in Fig. 7.4 and is sometimes referred as covering the Sargasso Sea. The flow is so slow and diffuse that it is hard to distinguish specific currents, although the *Canary Current* is recognized flowing to the south off the coast of North Africa.

7.331 Gulf Stream system

A more recent interpretation of the circulation, presented by Worthington (1976), questions several aspects of the classical pattern. His argument is based chiefly on a study of water mass characteristics, and the circulation which he proposes differs from the earlier pattern in two major respects. He argues that the circulation must be confined almost entirely to the north-western half of the North Atlantic, i.e north-west of a line from Iceland to the mouth of the Amazon River, rather than extending over the whole ocean as in the previously described circulation pattern; he also suggests that the western circulation consists of two gyres, not one. The main reason for the limitation to the north-west half of the ocean is that (diluted) Mediterranean outflow water persists as a saline wedge extending from the coast of Europe and North Africa westward to Bermuda or further west (e.g. Fig. 7.12(b)) and from depths of a few hundred metres to more than 2000 m and that a southward flow through this mass is extremely unlikely. It is not suggested that this water itself forms a barrier to flow but that, in view of the small rate of supply of Mediterranean water (about 1 Sv), its continued existence is evidence that there can be no significant flow from north to south (of water of different characteristics) at the depths occupied by this water mass in the eastern North Atlantic. The *Gulf Stream System*, according to Worthington, then consists of a clockwise gyre of which the northern edge is the Florida Current and the Gulf Stream flowing north-east or east with transports of up to 150 Sv. Most of this recirculates to the south and west back into the Gulf Stream within the western half of the ocean. The main part of this gyre lies between 30° to 40° N and 40° to 80° W with about 30 Sv of the recirculation penetrating further south, through the Antilles and the Caribbean, to form the Florida Current. The second and smaller circulation, the "Northern Gyre", lies between about 38° to 52° N and 34° to 47° W, i.e. to the north-east of the Gulf Stream System, and has a maximum transport on its west side of about 74 Sv. This means that any flow to the east across the North Atlantic is not a direct continuation of the Gulf Stream itself. Worthington uses the term "North Atlantic Current" for the strong flow on the west side of the Northern Gyre. He mentions the possibility of irregular exchange of water between the two gyres.

It is admitted that some surface layer flow may occur in the eastern North Atlantic above the Mediterranean Water but Worthington finds little evidence for a North Equatorial Current as a significant volume transport contributing to the Gulf Stream System although he states that there are very few direct current measurements in the eastern tropical North Atlantic to confirm this.

The circulation paths are based on a study of the water mass characteristics for the whole, while the transport values are based on both geostrophic calculations and direct current measurements in the north-western North Atlantic. One of the results is a series of estimates of vertical flows by means of a so-called "box model" in which the ocean is divided from bottom to surface into five layers by isothermal surfaces at 4°, 7°, 12° and 17°C. Circulation patterns and horizontal transports are estimated for each of these layers and the vertical transports are determined by assuming continuity of volume and of water characteristics. This box model leads to an average residence time for water in the North Atlantic of 241 years, which makes it one of the youngest of the oceans. (Note that as defined in Chapter 5, "residence time" is not necessarily the same as "age".)

It must be pointed out that Worthington's interpretation is regarded as controversial. Specifically, Clarke, Hill, Reiniger and Warren (1980), from a multi-ship study off Newfoundland, have presented arguments against Worthington's two main points of difference from the classical circulation pattern. They point out that Worthington's two-gyre hypothesis involves a rejection of the basic physics of geostrophic circulation and that lateral mixing can account for the water property differences which Worthington explains by his two gyres. Clarke *et al.* do recognize an eddy to the east of the Grand Banks (also observed earlier by Mann, 1967) but consider this to be a minor feature and that the North Atlantic Current is definitely a continuation of the Gulf Stream and that there is only one main gyre in the North Atlantic. They do agree that if the Mediterranean influenced water mass in the North Atlantic were maintained by *advection* (flow), Worthington's argument about no southward flow in the eastern North Atlantic would be valid. However, they point out that the Mediterranean outflow is small and that several studies have shown that the saline water mass in the North Atlantic could be maintained by lateral *diffusion* to the west from the high salinity Mediterranean outflow water concentrated along the eastern shore of the Atlantic, even in the presence of an ocean-wide gyre with slow southward flow in the east as envisaged by the classical picture of the North Atlantic circulation.

It is by such advancement of hypotheses or models, followed by critical study which may either support them or suggest revision, that our understanding of the ocean circulation advances. Accounts of the development of ideas about the Gulf Stream have been presented by Stommel (1965) and by Fofonoff (1981).

The salient feature of the Gulf Stream System in either the classical pattern or that of Worthington is the fast, concentrated north-eastward flow of the Florida Current and the Gulf Stream on the west side (westward intensification) contrasting with the broad and less well-defined south and westward flow over the rest of the ocean. The Gulf Stream is sometimes misrepresented as an individual current of warm water distinct from its surroundings, like a "river in the ocean". It is better thought of as the rapidly moving western edge of the Sargasso Sea, the 700- to 800-m deep pool of warm water which forms the upper layer of most of the North Atlantic Ocean. The Florida Current, which precedes the Gulf Stream, flows over the continental slope but the Gulf Stream itself is over deep water. The speeds of up to 250 cm/s (9 km/h) in the Gulf Stream are among the highest found in ocean currents. Radar altimeter measurements from *Seasat* for 3 months in 1978 (Cheney and Marsh, 1981) gave an average speed (calculated geostrophically from sea surface slope) of 150 cm/s, average width of 115 km and average change of surface elevation across the Stream of 140 cm, corresponding to a surface slope of 1.2×10^{-5}. The Gulf Stream is usually sharply defined on its north-west side, but much less so on the south-east toward the centre of the gyre. Due to the small scale of Fig. 7.4 the Gulf Stream is shown as a series of straight arrows directed to the north-east. However, when examined in detail it can perhaps best be described as consisting of a series of filaments of current which are usually sinuous or meandering. While one can indicate on a chart the general region where these filaments which comprise the Stream are likely to be found, it is not possible to predict where they will be individually at any particular time. Some of these meanders were recorded in detail during an unusual (at that time) oceanographic survey in 1950, called "Operation Cabot" (Fugilister and Worthington, 1951), when six oceanographic ships investigated a part of the Gulf Stream region simultaneously for 3 weeks. In particular, one meander developed to the extent that it broke off to form an individual eddy while the filament of current closed up behind it. More will be said about such eddies and similar features in the ocean in Section 7.334.

The North Atlantic Current is sometimes described as consisting of several branches or filaments of current forming a continuation of the Gulf Stream. Fugilister (1955) took data for the region between 75° W and 25° W for several cruises within a short period of time in 1953 and showed three significantly different interpretations of the same data. These interpretations varied from a single rather tortuous stream to a large number of fairly straight filaments. This exercise demonstrated the deficiencies in our observational knowledge of this region. The density of observations which it is practical to make from ships is still low and such observations are not sufficiently simultaneous to give an adequate picture of a complex region. However, observations from aircraft and from satellites can in many cases delineate temperature boundaries in the Stream and neighbouring waters and are helping to improve our knowledge of the Stream's location and of its changes with time.

Between the Gulf Stream and the shore of North America there is a south-westward-flowing coastal current with an elongated counter-clockwise gyre between it and the Stream. The exact mechanism which maintains this gyre has been much discussed as a problem in dynamical oceanography. The coastal current is partly supplied from the Labrador Current which flows south out of the Labrador Sea (Fig. 7.19) and round Newfoundland. The presence of this southward-flowing, cold, low-salinity Labrador Sea water in proximity to the warm, saline waters of the Gulf Stream gives rise to a very complicated oceanographic situation in this western part of the North Atlantic.

7.332 Gulf Stream volume transport

According to the 1976 review of the North Atlantic circulation by Worthington, the average transport of the Florida Current (off Florida) is 30 Sv (with variations from 15 to 38 Sv according to other studies using the electromagnetic method). The transport increases north-east to 85 Sv off Cape Hatteras and to 150 Sv by 65° W and then decreases eastward to 37 Sv at 40° W.

The increase west of 65° W is supplied by recirculation to the south-west from east of this longitude including water from the Antilles Current. A significant part of the increase may be in the form of Gulf Stream rings which could account for some 40 Sv.

One feature of the circulation which was observed directly with Swallow floats (Swallow and Worthington, 1957, 1961) was that underneath the north-east-ward-flowing Gulf Stream was a south-westward-flowing countercurrent with speeds of 9 to 18 cm/s. Earlier studies using the geostrophic relationship had suggested the possibility of such a current but the different interpretations were not consistent and to many oceanographers the idea of a significant current close to the bottom was not acceptable. However, Stommel (1965) had developed a theory of the thermohaline circulation from which there was good reason to expect such a southward deep current. The 1957 and 1961 measurements with Swallow floats demonstrated that in the region about 33° N, 75° W it certainly did exist at those times. Subsequent measurements (see Fofonoff, 1981) have shown flows varying from 2 to 50 Sv with a mean of 16 Sv. Wüst computed that a similar southward flow must occur in the western South Atlantic. There is some indication of a southward flow below the Kuroshio on the west side of the North Pacific but the measurements are few and of short duration and do not really provide convincing evidence.

7.333 Gulf Stream temperature and salinity distributions

Figure 7.6 shows the distribution of temperature and salinity in a section across the Gulf Stream. The Stream itself is associated for dynamic reasons with the steeply sloping isotherms and isohalines in a relatively narrow band of about 120 km width as shown. The fact that the water to the left of the current is cold compared with that at the same depths to the right, and the steepness of the isotherms in such a section, gave rise to the term "cold wall" to describe the water to the left of the Stream. Although the term is striking, it must be remembered that the real slope of the isotherms in the sea here is only of the order of 1 in 200. This is certainly very steep for the slope of a property surface in the sea but it is not much for a wall! It is clear from the sections that the Gulf Stream is not an individual flow of warm water and is hardly distinguishable from the Sargasso Sea to its east. One other feature is the wide spacing between the 16° and 20° isotherms over most of the section (see also Fig. 7.8(b)). A body of so-called *18° Water* has been a permanent feature of the west side of the Sargasso Sea since it was first observed from the *Challenger* in 1873, and its properties are very consistent within a few tenths of a degree on either side of 18° C and with a salinity of 36.4 to 36.6°/$_{oo}$. If it is winter-formed water the remarkable feature is the consistency of its properties, particularly temperature considering the variation of air temperatures in winter. Estimates of the annual rate of formation of this water vary from about 50% of its

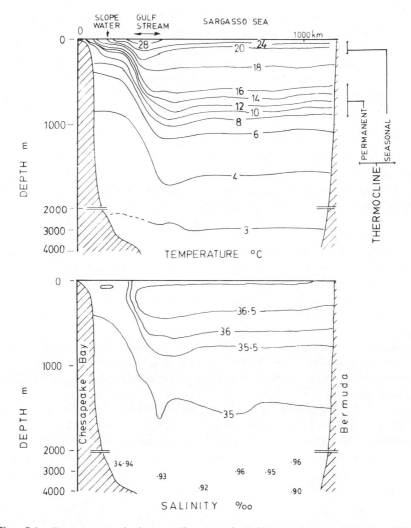

FIG. 7.6.   *Temperature and salinity profiles across the Gulf Stream, August–September 1932.*

average volume to zero. The rate of formation is calculated from the heat loss expected in winter and the differences in rate of formation are due to the use of different eddy diffusion coefficients (Chapter 5), particularly for evaporation, by different workers. As Worthington (1976) points out, "Of the problems in ocean circulation, that of the amount and manner of water-mass formation is the least tractable".

For many years, information on the Gulf Stream area of the North Atlantic has been collected by the U.S. Naval Oceanographic Office and later by the

National Oceanic and Atmospheric Administration from a variety of sources such as government and commercial ships, from aircraft using radiation thermometers, photography and radar, and from satellites. Since 1966 a digest of this information has been published monthly with sea temperatures, Gulf Stream position, and articles on special features of interest in this region ("*gulfstream*", N.O.A.A., U.S. Department of Commerce).

7.334 Gulf Stream rings and ocean eddies

The cold or cyclonic eddy observed in "Operation Cabot" to form from a meander of the Gulf Stream drew attention to this process and since about 1970 much more information has been obtained. Fugilister suggested the name *Gulf Stream Rings* for these features which entrain colder Slope Water into the warm Sargasso Sea. Parker (1971) illustrated this process of formation as shown in Fig. 7.7. In this example, the result is an anticlockwise (cyclonic) rotating ring of cool, Slope Water in the warm Sargasso Sea. Alternatively, a meander may start on the north side of the Gulf Stream and result in a clockwise rotating ring of warm Sargasso Water in the cool Slope Water. Although first observed in the Gulf Stream region such rings are now known to be general features of the ocean circulation, the sense of rotation for the same type of ring being opposite in the two hemispheres to satisfy the geostrophic relationship. A cold ring is the oceanic equivalent of an atmospheric low-pressure system, i.e. a cyclone, while a warm ring is the equivalent of a high-pressure system, i.e. an anticyclone. For this reason, the sense of rotation of a cold ring is often referred to as *cyclonic* (anticlockwise in the northern hemisphere, clockwise in the southern) and for a warm ring as *anti-cyclonic* (clockwise in the northern hemisphere, anticlockwise in the southern). These rings are 150 to 300 km in diameter and some 3000 m in vertical dimension and have lifetimes of up to 2 years. At any time there may

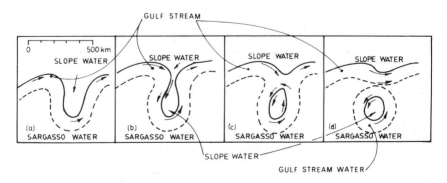

FIG. 7.7.   *Diagram of Gulf Stream ring formation.*

be three or more warm-core rings north of the Stream and eight to fifteen cold-core rings south of it in the area west of 50°W and north of about 30°N. After formation the rings move in a south or south-westerly direction at speeds of a few kilometres per day and eventually merge with the Gulf Stream again (e.g. see Richardson, 1980). In a newly formed cold ring the thermocline is about 500 m higher than in the surrounding water and, in addition, it carries both the chemical characteristics and the biological populations of the Slope Water into the Sargasso Sea area (Ring Group, 1981).

In Fig. 7.8(a) are shown, from a study by Richardson, Cheney and Worthington (1978), the positions of nine cold and three warm rings in the

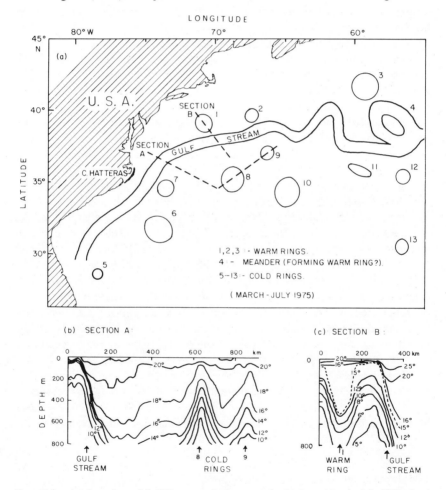

FIG. 7.8.   (a) Locations of Gulf Stream and of warm and cold rings, March–July 1975, (b, c) Vertical temperature sections along lines A and B in (a) showing Gulf Stream and cold and warm ring structures.

spring of 1973, obtained from four XBT and CTD surveys together with other data centre information, including satellite observations (infra-red). There was also a meander at 62°W which had formed when a warm ring coalesced with the Gulf Stream; it later formed a new warm ring. At 57°W there was a meander/ring which continued during the period of study. In Fig. 7.8(a) are shown two lines, A and B, which show the positions of two vertical temperature sections reproduced in Figs. 7.8(b) and (c). In section A, the Gulf Stream appears at the left where the isotherms slope down steeply to the east and two cold rings are shown at 8 and 9 where lower temperature isotherms rise up toward the surface underneath the thermocline. In section B, the warm ring 1 appears at the left where the isotherms dip down to a maximum depth at the 100 km distance, while the Gulf Stream is at the right.

Other features seen in section A in Fig. 7.8(b) are the 18° water, and evidence of the Gulf Stream recirculation to the west revealed by the isotherms (which approximate isopycnals here) sloping up to the east at 300 km distance. The slope of the isotherms for ring 8 shows the cyclonic direction of rotation, to the south at the west side of 8 and to the north at its east side.

Seasat altimeter measurements in 1978 (Cheney and Marsh, 1981) indicated that the surface level at the centre of cold rings was up to 95 cm deeper than at the edge while in warm rings it was up to 75 cm higher at the centre than at the edge.

A now accepted feature of the open ocean circulation about which there have been hints since the 1930s or even earlier (see the article by Wunsch, 1981) is that embedded in the general large-scale circulation there are large numbers of eddies, some circular, some elongated. They have been studied in the North Atlantic since about 1960, most intensively since about 1970, e.g. in the Soviet POLYGON-70 study and the U.S. MODE-1 study in 1973 (MODE Group, 1978) which were precursors to the 1978 international POLYMODE study. The eddies are typically tens to hundreds of kilometres across and often extend from the surface to the bottom. The mode of formation of the Gulf Stream rings and others near major currents is clear (although the reasons for the meanders starting are not obvious) but the origin of the mid-ocean eddies is not yet understood. The ocean eddies have been likened to the storms and weather systems of the atmosphere, the smaller size and speed of motion of ocean eddies being related to the greater density of water than air. In spite of these characteristics, ocean eddies have been observed to persist over periods of many months. An excellent acount of Ocean Eddies is presented in the Spring 1976 edition of *Oceanus* (MODE Group, 1976).

7.34 Equatorial Atlantic circulation

As the major features common to the oceanic equatorial circulations are most clearly developed in the Pacific Ocean they will be included in the description of that ocean later in this chapter and only a brief description will be given here for the Atlantic. In the Equatorial Atlantic surface layer (Fig. 7.4), according to the classical picture of the circulation there are, from north to south, the North Equatorial Current flowing to the west, then the eastward-flowing saline North Equatorial Countercurrent between 9° and 5°N and then the South Equatorial Current from the eastern South Atlantic, part of which current crosses the equator to flow west and north along the coast of Brazil. The North Equatorial Countercurrent is supplied from both the North and South Equatorial Currents; it often disappears from the surface in the west during the first half of the year. Below the surface and flowing to the east along the equator is the Equatorial Undercurrent. The South Equatorial Current supplies the high salinity water for the core of the Undercurrent. In the west this latter current is variable but from 30°W to near the African coast it is a permanent feature of the equatorial current system. Its core is at 60 to 100 m depth and with speeds to over 100 cm/s it has a volume transport from 15 to 35 Sv. The characteristics of equatorial undercurrents will be described in detail in the Pacific Ocean section.

7.35 Atlantic Ocean water masses

When studying the water mass characteristics of the Atlantic Ocean we are fortunate in having a considerable amount of information and in particular the data from two extensive expeditions, both of which covered the greater part of the area systematically and in a relatively short time. The two expeditions were the German *Meteor* expedition during 1925 to 1927, and the International Geophysical Year studies of 1957–58 carried out by Woods Hole Oceanographic Institution of the United States with cooperation from the National Institute of Oceanography of Great Britain (Fugilister, 1960). The second of these studies was deliberately arranged to cover the South and Equatorial Atlantic along the same lines of stations as those occupied by the *Meteor* in order to obtain a direct comparison of the distribution of water properties after the interval of 30 years. One of the immediate results of the comparison of the two sets of data was the recognition that the distributions of temperature, salinity and dissolved oxygen were almost identical. This result was very comforting to oceanographers who have made a practice of assuming that the state of affairs which they observe in the ocean is a reasonably steady one, at least when averaged over a year to eliminate seasonal changes.

It should be noted that the spacing between stations in the above surveys was too large to resolve the mesoscale eddies mentioned earlier. These eddies are a source of shorter time-scale changes which, when averaged over a sufficient time, do not alter the mean large-scale circulation studied by the surveys.

7.351 Atlantic Ocean upper waters

In the surface layer of the Atlantic, the salient temperature characteristic is that it is high at low latitudes and decreases to higher latitudes (Figs. 4.3, 4.4). Between the equator and about 20°N the temperature ranges between 25° and 28°C with little seasonal change (cf. the Pacific Ocean, Fig. 4.7). In the South Atlantic, south of the equator to 40°S, there is a seasonal change of temperature of about 5K between winter and summer; at higher latitudes the seasonal range of temperature decreases. In the North Atlantic the difference between winter and summer temperatures rises to 10K at 40°N and then decreases at high latitudes. The salinity does not show any significant seasonal change (except close to ice). There is a minimum value of 35‰ just north of the equator and then maxima of 37.3‰ just north of the equator and then maxima of 37.3‰ in the tropics at about 20°N and S. From here the salinity decreases to 34‰ or less at high latitudes.

There is a marked difference in the surface water characteristics between the west (American) side and east (European–African) sides. In the North Atlantic there is a difference of about 25C° in the sea-surface temperature between Florida and Labrador, compared with only about 10K between the same latitudes on the east side (North Africa to Scotland). In salinity there is a south–north difference of about $3^o/_{oo}$ on the west compared with $1.5^o/_{oo}$ on the east. These differences between west and east are clearly associated with the differences in currents. In the west, there is the contrast between the warm, saline Florida Current in the south and the cold, low salinity Labrador Current in the north whereas on the east side there is just the slow, diffuse southward flow of the North Atlantic gyre with no contrasting water masses.

In the South Atlantic, the effect of the Brazil Current is present near the equator in contrast to the Falkland Current in the south but the differences in properties are not as marked as in the North Atlantic. On the east side of the ocean, low temperatures immediately off the coast of South Africa are due to upwelling of subsurface waters. Low salinities in this region are also characteristic of this water but the low salinities farther north, off tropical Africa, are due to river runoff.

7.352 Atlantic Ocean deep-water masses and circulation

The main features of the deep-water characteristics and circulation of the Atlantic are well shown in Fig. 7.9 which is based on the GEOSECS data for the Atlantic collected in 1972–73 (Bainbridge, 1976). The four sections show the distributions of potential temperature, salinity, potential density and dissolved oxygen along the western trough of the Atlantic Ocean, i.e. between the American continent and the Mid-Atlantic Ridge. The temperature section shows clearly that the largest variations both horizontally and vertically are in the upper layer, and emphasizes the statement made earlier that vertical

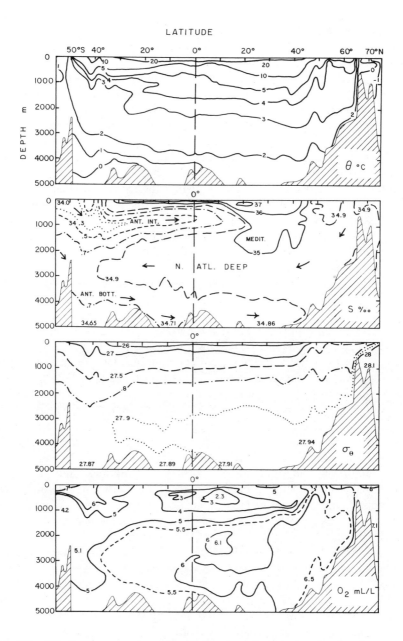

FIG. 7.9.   *Atlantic Ocean—south–north vertical sections of water properties along the western trough (data from GEOSECS Atlas, 1976).*

property gradients in the sea are generally much greater than horizontal ones.

The Antarctic Intermediate Water from south of the Polar Frontal Zone is evident to some extent as a low-temperature tongue centred at 1000 m, but is much clearer as a low salinity tongue extending to 25°N. It is also apparent, although less clearly, by an oxygen maximum at about 800 m depth to about 10°S. Below this and flowing along the bottom to the north is the cold Antarctic Bottom Water which can be traced by its low salinity to about 45°N.

Between the Intermediate Water above and the Bottom Water below there is the great bulk of the Deep Water extending from the North Atlantic to the Southern Ocean and most evident in the salinity and oxygen sections. (It also appears to be shown by the σ_θ section but this is an artifact which will be discussed later in this section.) For a long time the accepted origin of this Deep Water was in the Labrador and Irminger Seas off the southern tip of Greenland where winter cooling of relatively saline water was considered to cause it to sink to considerable depths and then spread south. This suggestion was made by Nansen (1912), and Dietrich (1969) presented data from 1958 which apparently showed this process happening. In the region south of the southern tip of Greenland the water properties in late winter were almost uniform from the surface to near bottom. For instance, the temperatures from the surface to 3000 m ranged only between 3° and 3.25°C in an area over 150 km across. From this column, a tongue of 3° to 3.25°C water extended south at depths of 2000 to 2500 m. This water had a high oxygen content; even at 2000 m depth it was still 90% of that at the surface. The explanation for this situation was that the deep water was actually forming at the time, being cooled at the surface and sinking rapidly to 2000 to 3000 m depth, and then continuing south at its density level as North Atlantic Deep Water.

However, no calculations were made of the volume of water so formed and, as accumulating data indicated that this process did not take place every winter, attention was directed to other possible sources or mechanisms. As a result both of studies of water property distributions and of current measurements, following a suggestion made by Cooper (1955), it is now recognized (e.g. Crease, 1965) that the major source of North Atlantic Deep Water (up to 80%) is the Norwegian Sea from which Deep Water flows over the sills between Scotland, Iceland and Greenland and cascades into the depths of the Atlantic. Lee and Ellett (1965) suggested the following scheme. Over the sill between Scotland and Iceland, Norwegian Sea Deep Water (-0.5°C and 34.86 ‰) flows south and mixes with North Atlantic (Central) Water (9°C and 35.33 ‰) above it to form the *Northeast Atlantic Deep Water* (2.5°C and 35.03 ‰). Some of this Northeast Atlantic Deep Water continues south in the eastern basin of the North Atlantic over bottom water of Antarctic origin. The remainder of the Northeast Atlantic Deep Water flows west across the Mid-Atlantic Ridge into the western basin, below the Labrador Sea water in the north (3.4°C and 34.89 ‰) but above the *Northwest Atlantic Bottom Water*

(1.0°C and 34.91 $^o/_{oo}$). This latter water originates in an overflow of Norwegian Sea Deep Water across the sill between Iceland and Greenland. This flows south-west along the continental slope of Greenland, picking up first some modified Labrador Sea Water and, when deeper, some Northeast Atlantic Deep Water, the proportions probably varying from time to time. This mixed water descends into the western basin of the North Atlantic to form the Northwest Atlantic Bottom Water. The North Atlantic Deep Water which flows south across the equator therefore has two components, the Northeast Atlantic Deep Water and the Northwest Atlantic Bottom Water (see Warren, 1981 for a fuller review).

One other point to mention is that both current meter measurements and water property distributions suggest that the overflows between Scotland, Iceland and Greenland are not continuous but probably occur in pulses—the dynamic reasons are not known.

The density characteristic plotted in Fig. 7.9 is σ_θ for the following reason. At shallow depths, the difference between T and θ, and so between σ_t and σ_θ, is small and either may be used to describe the density distribution. In deeper water, the difference between T and θ becomes significant and σ_θ is preferred because the effect of pressure on the temperature of the water has been removed. In the upper 1000 m or so, σ_θ follows θ rather than S, e.g. the effect of the low salinity of the Antarctic Intermediate Water is not apparent in the σ_θ section. However, in the deeper water, the effect of salinity is seen in the relation of σ_θ to the North Atlantic Deep salinity maximum combined with decreasing temperature (Lynn and Reid, 1968; Reid and Lynn, 1971).

The decrease of σ_θ with increasing depth below about 4000 m was earlier thought to indicate that the bottom layer was gravitationally unstable because it appeared that denser water was above less dense. This is not really the case— it is an artifact of using σ_θ, i.e. $\sigma_{\theta, S, 0}$, the value for pressure = 0. Because the thermal expansion and compressibility of sea-water vary with temperature (in particular, colder water is more compressible), the vertical rate of change of σ_θ is not a good measure of stability when considered over finite depth ranges. A better estimate is obtained by using a value of σ for a higher pressure, e.g. σ_4 = $\sigma_{\theta, S, 4}$ for a pressure of 40,000 kPa (= 4000 dbar) which is close to the pressure at 4000 m depth. If this is done for the deep Atlantic, the value of σ_4 increases to the bottom, as shown in Fig. 7.10.

The bottom waters in the western and eastern basins of the Atlantic have noticeably different properties in the South Atlantic because of the barrier of the Mid-Atlantic Ridge. Figure 7.11(a) shows transverse sections of tempera-ture and salinity across the ocean at 16°S to illustrate this. On the west side the Antarctic Bottom Water has temperatures down to 0.4° C whereas on the east side the minimum is 2.4°C. The Antarctic Bottom Water is prevented from flowing directly into the eastern basin by the Walfish Ridge with a sill depth of less than 3500 m extending from Africa to the Mid-Atlantic Ridge. The salinity

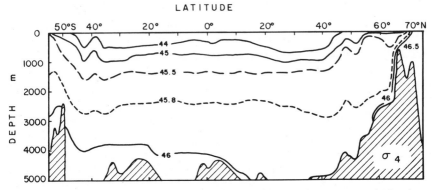

FIG. 7.10.   Atlantic Ocean—south–north vertical section of σ_4 along the western trough (data from GEOSECS Atlas; 1976).

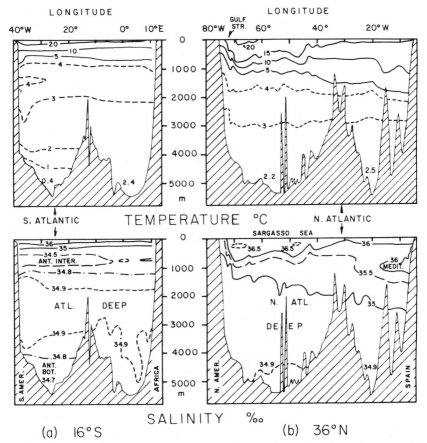

FIG. 7.11.   West–east sections for temperature and salinity across the Atlantic Ocean at 16° S and 36° N.

difference between the two sides is less remarkable, $34.7\%_{oo}$ on the west compared to $34.9\%_{oo}$ on the east. It is interesting to note that the densities on either side of the Ridge are identical, the temperature and salinity differences being in such directions as to cancel their effects on density. Other water masses evident in the figure are the Antarctic Intermediate Water with its low salinity core at about 800 m and the more saline North Atlantic Deep Water with its higher salinity core at about 2500 m.

Farther north at $36°N$ (Fig. 7.11(b)) the temperature difference has fallen to only 0.3K and the salinities are almost identical. The influence of the outflow of the Mediterranean Water is evident in the salinity section with its core at about 1000 m, but is less evident in the temperature section because the water has settled to a level determined by its density which is determined mainly by temperature at this depth. At the west the steep slopes of the isotherms and isohalines associated with the Gulf Stream are also evident in the upper water.

Figure 7.12 shows horizontal sections of temperature (T) and salinity at 1000 m depth for the North Atlantic (Wüst and Defant, 1936) showing how this water spreads west and north across most of the North Atlantic.

7.353 T–S characteristics of subsurface waters of the Atlantic Ocean

The average T–S characteristics of the waters below the surface layers are shown in Fig. 7.14. The upper waters below the surface layer appear on this diagram as the *Atlantic Central Waters*, *North* and *South*, extending to depths of 300 m on either side of the equator and deepening to 600 to 900 m at mid-latitudes and then getting somewhat shallower at high latitudes. Both of the Central Water masses appear on the T–S diagram as straight lines extending from high temperature and salinity to lower temperature and salinity. At first sight, these Central Water Masses might appear to be examples of water masses produced by the vertical mixing of water types represented by the characteristics at the end of the lines (on the diagram). This, however, is believed not to be the case. Iselin (1939) pointed out that if one examines the winter T–S characteristics in the north–south horizontal direction at the surface over a range of latitudes near the subtropical convergences one finds the same T–S characteristics that one finds in the vertical below the surface in the Central Water Masses at lower latitudes. He then suggested that these water masses originate by sinking on the equatorward side of the subtropical convergences. The cooler, less saline water sinks at higher latitudes while the warmer, more saline water sinks at lower latitudes and flows equatorward above the cooler water. In the South Atlantic, the Central Water terminates at depth where it merges into the well-defined Antarctic Intermediate Water. In the North Atlantic, the Arctic Intermediate Water is a much less significant water body and the Central Water merges into this at high latitudes and into the Mediterranean Water at lower latitudes.

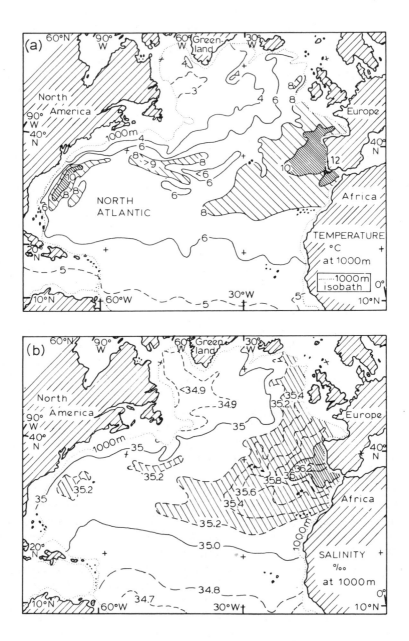

FIG. 7.12.   *North Atlantic: horizontal sections at 1000 m depth of (b) temperature, (a) salinity,*
to show the spreading of Mediterranean Water.

The effect of the Mediterranean Water can be seen in the T–S diagram (Fig. 7.14(a)) as a salinity maximum above the Deep Water maximum. The Mediterranean maximum is very marked for waters of the eastern North Atlantic but is less conspicuous in the west and in the south as it gradually loses its characteristic properties by mixing with the waters above and below it. However, by careful analysis using the core method, the Mediterranean Water can be traced down through the South Atlantic.

Two points must be mentioned about Figs. 7.13 and 7.14. The first is that the lines indicate mean values for large areas; the individual points scatter over a band whose width is about one-third of the separation of the two lines which represent the mean values for the South and the North Atlantic, i.e. a band of width equivalent to about 0.1 $^{\circ}/_{oo}$ on the diagram. (This is also true for the other T–S curves in Figs. 7.13 and 7.14.) To better illustrate this scatter we present, as an example, Fig. 7.15 which shows all available data points for a 10° square, i.e. 10° latitude by 10° longitude, for the tropical Pacific south of Hawaii. In this figure, the T–S relationship is expressed by three lines. The centre line represents the mean T–S curve for that square while the bracketing lines represent one standard deviation in salinity on either side of the mean for all data within that square. Thus we can examine not only the distribution of the mean property-to-property relationship but also the degree of variability in this relationship. This variability is usually greatest at the upper, or warm, parts of the curves, e.g. the Tropical Upper Water in Fig. 7.15. It is caused by changes at the sea surface which alter both the temperature and salinity. In some regions, lateral meanders of boundaries between water masses also cause variability even below the surface as in Fig. 7.15 where both the North Pacific Intermediate and the Pacific Equatorial Waters are found within the same square.

The second point is that Figs. 7.13 and 7.14 present only the salient features of the T–S curves averaged over large oceanic areas. When one looks at the T–S curves for local regions, considerable variations from the mean curves are found. To illustrate this, the T–S curves for all 5° squares for the North Atlantic are plotted on a map in Fig. 7.16 (Emery and Dewar, 1982). In this display, the considerable variety of local T–S curves is apparent and one can easily trace the characteristic temperature and salinity values associated with individual water masses. Strong variations are seen over the entire temperature range at 40° to 50° N off the east coast of North America. These reflect meanders in the Gulf Stream system, carrying very different water masses north and south through the 5° squares. At the higher latitudes (above 50° N) in the western North Atlantic, the variability is largest at low temperatures as here the sea surface is coldest.

Considering the local mean T–S curves in Fig. 7.16 one can follow the transitions of various maxima and minima. At about 2° C the salinity minimum of the Antarctic Intermediate Water is well marked in the southern

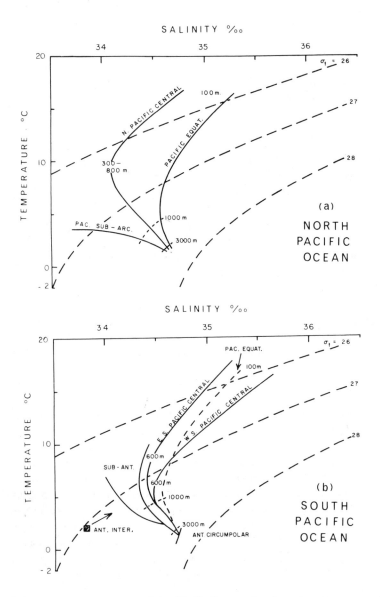

FIG. 7.13. *Average temperature–salinity (T–S) diagrams for the main water masses of the Pacific Ocean.*

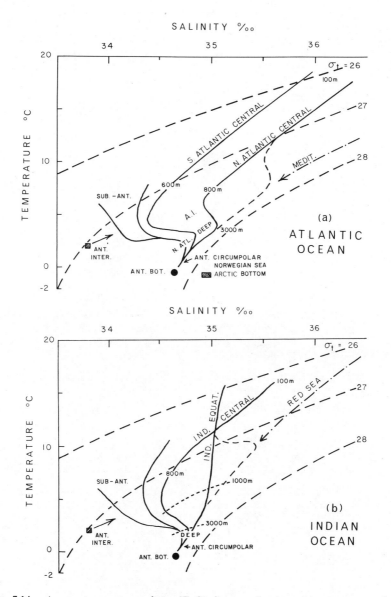

FIG. 7.14. *Average temperature–salinity (T–S) diagrams for the main water masses of the Atlantic and Indian Oceans.*

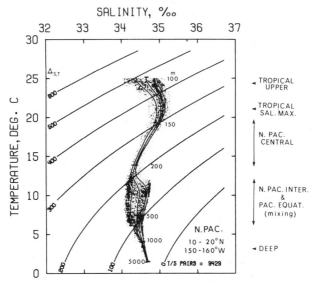

FIG. 7.15.   *Example of T–S "scatter plot" for all data within a 10° square with mean T–S curve (centre line) and curves for one standard deviation in salinity on either side.*

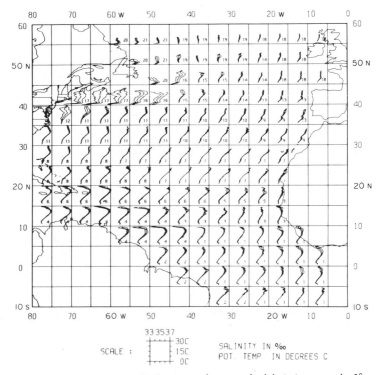

FIG. 7.16.   *Atlantic Ocean: mean T–S curves and one standard deviation curves by 5° squares.*

latitudes. This minimum erodes to the north, losing its character by about 30° N. In the south, an almost straight line links the Antarctic Intermediate Water with the salinity maximum ($\sim 36\,^{\circ}/_{\circ\circ}$) of the upper waters. To the north, these maximum values are found at the warmest temperatures.

In northern latitudes, the temperature range is limited to values below 15° C. As has been mentioned, in these regions the coldest temperatures are frequently at the top of the water column, emphasizing the significant role of salinity in determining density in high latitudes. These regions are marked by a cold salinity maximum. In contrast, the waters in the central North Atlantic appear almost isohaline over the entire temperature range.

At mid-latitudes, off the Strait of Gibraltar, the saline outflow from the Mediterranean leads to a salinity maximum at mid-depth at about 10° C. From the sharp bend in the T–S curves one can trace this maximum as it spreads north, west and south. As has been discussed, this water mixes with those above and below as it spreads, eroding the salinity maximum.

The great variety in T–S curve shapes shown in Fig. 7.16 (prepared in 1980) compared with those in Fig. 7.14 (prepared before 1942) is partly because the latter were intended primarily as a synopsis of the main water mass features for large areas but is chiefly because there are so many more data available now, making it possible to divide the ocean into relatively small areas (Fig. 7.16) but still have sufficient data within each for the T–S curves to be significant. Also, computer processing makes possible the examination of these large quantities of data which it would be impractical to analyse with "paper and pencil" methods.

It is left to.the reader to study Fig. 7.16 further to identify other features of the North Atlantic water mass characteristics and their spatial distribution, bearing in mind that most of the length of each T–S curve represents only the upper 500 to 1000 m of the water column as indicated, for example, by the depth marks in Fig. 7.15. It may also be noted that the paper by Emery and Dewar (1981) also contains maps of mean and standard deviation T, z and S, z profiles for the same 5° squares.

7.354 Mixing mechanisms

In recent years considerable attention has been devoted to the possible mechanisms causing mixing in a situation such as where the Mediterranean water flows out from the Strait of Gibraltar into the Atlantic, particularly in conditions when the turbulence level is low. The essentials of the situation are that warm, saline (Mediterranean) water lies over cooler, less saline (Atlantic) water (Fig. 7.11(b)). In this case the upper water is a little less dense than the lower, so that the two layers are in stable equilibrium. Because of the temperature and salinity differences between the layers, heat and salt will both diffuse down across the interface between the layers. If we assume that the vertical turbulent mixing across the interface can be neglected in this case, the transfer of properties will be by molecular processes. Because the rate of molecular diffusion for heat is about 100 times faster than for salt, the upper layer loses heat and the lower gains it faster than the upper loses salt and the lower gains it. In consequence the upper water becomes denser and tends to sink while the lower becomes less dense and tends to rise. In laboratory experiments with such layers the water movements occur in small columns or "fingers", a few millimetres across, and result in mixing between the upper and lower layers. The mechanism is referred to a *salt fingering* or, because it is a consequence of the two diffusion rates being different, as *double diffusion*. A very interesting feature is that the fingers only penetrate a limited distance beyond the original interface and the subsequent lateral mixing gives rise to a uniform layer. Then the process may start again at the two interfaces which are now present, and eventually a number of layers develop with sharply defined interfaces in terms of temperature and salinity. In the case where the mean temperature and salt gradients oppose each other, this process is suggested as a possible cause for the sharp steps which are often found on CTD traces, indicating clearly defined homogeneous layers of thickness scale from metres to tens of metres separated by thinner zones of sharp gradients of temperature and salinity.

The layers are on too small a thickness scale to be observed with bottle-cast sampling techniques and it was only when the continuous records of temperature and salinity with depth from CTD instruments became available that these layers were observed in the ocean. Step structures in temperature/depth and salinity/depth traces observed below the Mediterranean water in the Atlantic (Tait and Howe, 1968, 1971) were at first attributed to the double diffusion process but further study has suggested that in this area the steps may be due primarily to interleaving along density surfaces when water masses of different T–S characteristics come together laterally, the masses having different density/depth profiles. Such interleaving has been observed in the neighbourhood of the equator in the western Pacific (see Jarrige, 1973) where the saline South Pacific Subtropical Water moving north meets the less saline North Pacific Water. Salt fingers were first observed optically by Williams (1975). An account of fine-structure observations has been given by Federov (1976) and a recent review of fine-scale mixing processes by Turner (1981).

7.4 North Atlantic Adjacent Seas

7.41 Mediterranean Sea

The Mediterranean Sea (Fig. 7.17) is of interest in displaying some results of the interaction between the atmosphere and the sea. Due to the heat input and the large excess of evaporation over precipitation in the eastern part of the Sea

FIG. 7.17.   *Mediterranean, Black and Baltic Seas: surface circulation and regions of Intermediate and Deep Water formation for Mediterranean.*

the water in this region is characterized by high temperatures and salinities which are surpassed only in the Red Sea.

The Mediterranean is divided essentially into a western and an eastern basin by a sill of depth about 400 m extending from Sicily to the North African coast. The maximum depths are about 3400 m in the western basin and 4200 m in the eastern.

The two characteristic Mediterranean subsurface water masses are the Intermediate Water and the Deep Water. The former, called by Wust (1961) the *Levantine Intermediate Water*, is formed in the winter off the south coast of Turkey (Fig. 7.17) with a temperature of 15° C and a salinity of 39.1 %. It flows west at 200 to 600 m depth (Fig. 7.18) along the North African coast and out below the surface through the Strait of Gibraltar into the Atlantic (Figs. 7.11, 7.12). By the time it starts to flow down the continental slope its properties have been modified to 13° C and 37.3 % by mixing with overlying Atlantic

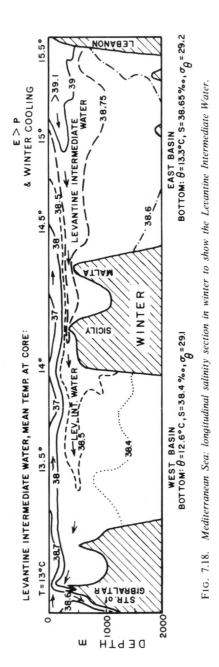

Fig. 7.18. *Mediterranean Sea: longitudinal salinity section in winter to show the Levantine Intermediate Water.*

water during its passage from the east and particularly through the Strait. The outflow below the surface is replaced by a surface inflow from the Atlantic (Fig. 5.2(a)). It should be noted that the water which emerges from the Mediterranean is the Intermediate Water and not directly water which has spilled or been displaced out of the deeper basins.

The *Deep* and *Bottom Waters* are formed at the northern edges of the basins in winter, chiefly off the Riviera in the western basin and in the southern Adriatic Sea in the eastern basin (Fig. 7.17). In the western basin the deep-water potential temperature is 12.6° C and the salinity 38.4 $\%_{oo}$, in the eastern basin the properties are 13.3° C and 38.65 $\%_{oo}$ with very high σ_θ values of up to 29.2. These values may be contrasted with those at 4000 m in the Atlantic of 2.4° C, 34.9 $\%_{oo}$ and a σ_θ of 27.8. The oxygen content of the deep water in these basins is fairly high, up to 4.7 mL/L in the western and 5.0 mL/L in the eastern, suggesting frequent replenishment by the considerable vertical convection which carries the winter-formed water down to the depths of the basins.

In January to March 1969, a six-ship operation off Toulon in the south of France observed this process in action for the first time. It took place in three phases. The preconditioning phase was when the winds from the north (Mistral) cooled the surface water until it was only just stable. Then a violent mixing phase occurred when stronger, dry winds blew for 6 days during which the water column was further cooled and was mixed to uniformity down to 1400 m depth. After this the water sank to depths of at least 2400 m and spread out, displacing upward the older deep water. (The maximum depth in this north-western basin is less than 2900 m.) A very interesting feature is that the horizontal extent of the formation area was only about 40 km across but nevertheless the volume of deep water formed (in a few days) was estimated to be equal to the outflow through the Strait of Gibraltar for about 6 weeks. (For a fuller account see Sankey, 1973.)

7.42 Black Sea

The oceanographic characteristics of the Black Sea (Fig. 7.17) which connects with the eastern Mediterranean are in direct contrast with those described above. The Black Sea has a maximum depth of 2240 m and connects with the Mediterranean through the narrow passages of the Bosphorus and the Dardanelles, which have depths of only 40 to 100 m. The Deep Water in the Black Sea has a salinity of only 22.3 $\%_{oo}$ and a potential temperature of 8.8° C. There is no dissolved oxygen below about 180 m; it is replaced by hydrogen sulphide whose concentration increases with depth.

There is evidence that over geological time the water in the Black Sea has varied from being fresh to moderately saline and it is not known whether the present deep-water salinity represents a steady state or not. There is considerable river runoff of fresh water into the Sea and, at most, a very limited influx of salt water from the Mediterranean. The currents through the Bosphorus and the Dardanelles consist of a surface outflow of low-salinity water, carrying the precipitation and river runoff, and a subsurface inflow of more saline water. The narrowness and shallowness of the passage results in considerable current speeds and current shear (variation with depth) and the consequent turbulence causes vertical mixing. The result is that the surface water which leaves the Black Sea with a salinity of about $16\,^o/_{oo}$ reaches the Mediterranean with its salinity increased to $30\,^o/_{oo}$, while the incoming subsurface water which leaves the Mediterranean with a salinity of $38.5\,^o/_{oo}$ has this reduced to 35 to $30\,^o/_{oo}$ by the time that it reaches the Black Sea. There is some difference of opinion as to how much saline water actually flows into the Black Sea. There is no doubt that there is a considerable flow from the Mediterranean into the Dardanelles but much of this saline water is mixed upward and carried out in the surface layer and so does not reach the Black Sea. This is an example of the estuarine type of circulation which will be discussed in Chapter 8. It is evident from the stagnant condition of the deep water of the Black Sea, as discussed in Chapter 5, that the volume of saline (and well-oxygenated) water which penetrates through the Bosphorus and sinks in the Sea must be small and insufficient to refresh the deep layers.

7.43 Baltic Sea

The Baltic Sea (Fig. 7.17), including the Gulf of Bothnia and the Gulf of Finland, shows some features which may be compared with those of the Mediterranean and Black Seas. In contrast to the Mediterranean the Baltic is shallow, with an average depth of only 100 m and a maximum of about 460 m. Like the Black Sea it has an excess of precipitation and river runoff over evaporation. In consequence the salinity is low, usually below $10\,^o/_{oo}$ at the surface and rising to only 10 to $16\,^o/_{oo}$ at 5° C in the deep water. The Baltic, however, does not exhibit extensive stagnation, possibly because of the moderate depths. The communication with the North Sea outside is through shallow passages with a sill depth of 18 m and much mixing occurs between the outflowing surface layer and the inflowing more saline water. As a result of this the average inflow of saline water into the Baltic is small. However, tides play a significant part in assisting the introduction of saline water. In addition, the stress due to the wind is sometimes sufficient to override the two-layer flow and cause unidirectional flow at all depths through the entrance passages. Neither of these processes plays any significant part in the flow into the Black Sea.

7.44 Norwegian and Greenland Seas

Bordering the North Atlantic are two adjacent seas of some significance, the Norwegian and Greenland Seas to the east of Greenland and the Labrador Sea and Baffin Bay area to the west (Fig. 7.19). The *Norwegian Current* is a

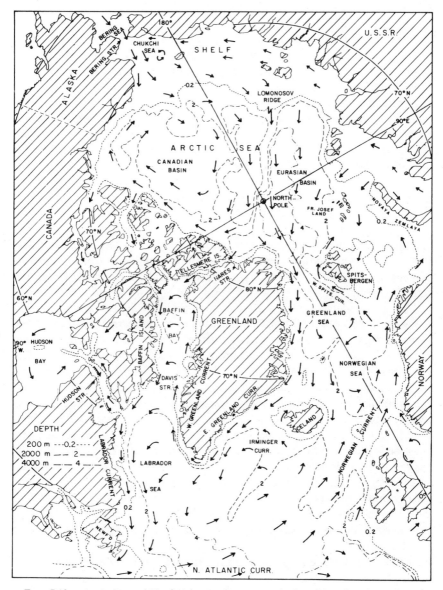

FIG. 7.19. *Arctic Sea and North Atlantic adjacent seas: bathymetry and surface currents.*

continuation of part of the North Atlantic Current which turns north and passes over the Wyville–Thompson Ridge between the Shetland and Faroe Islands into the Norwegian and Greenland Seas. Along the Greenland coast there is the south-westward-flowing *East Greenland Current* which is composed of the outflow from the Arctic Sea and some water from the Norwegian Current. The speeds in these two currents are up to 30 cm/s. They are upper-layer currents and the submarine ridges which extend from Greenland to Scotland with maximum sill depths of less than 1000 m prevent deeper Atlantic water from entering the Norwegian Sea and hence the Arctic. Between the two currents are gyral circulations in the Norwegian and Greenland Seas.

A rather curious feature is that apparently the subsurface water which enters the Arctic Sea from this area comes from the gyre of the Norwegian Sea to the south rather than from the Greenland Sea which is north of it and closer to the Arctic. This was shown by Metcalf (1960) from data obtained in the winters of 1951 to 1955. Characteristically, the water of the Greenland Gyre above 1500 m has properties of -1.1 to $-1.7°$ C and 34.86 to $34.90°/_{oo}$, while below this depth the water is almost isohaline at $34.92°/_{oo}$ with a temperature of $-1.1°$C or colder. The deep water of the Norwegian Gyre has the same salinity but its temperature is $-0.95°$C or warmer, properties which are similar to those of the Bottom Water of the Arctic Basin. This Norwegian Sea water is also found to the east and north of the Greenland Sea and apparently in some way forms a barrier to the passage of the colder Greenland Gyre water into the Arctic.

The deep waters of both the Norwegian and Greenland Seas have high oxygen contents of 6 to 7.5 mL/L indicating frequent formation. It had been assumed that winter cooling at the surface resulted in overturning which mixed the water from surface to bottom. However, Carmack and Aagaard (1973) have pointed out that although there is considerable evidence that the formation of *Greenland Sea Deep Water* (GSDW) is affected by the severity of the particular winter, when good winter measurements in this area became available there was no evidence of a homogeneous water column. Therefore, unless the formation takes place in a very small area, some other mechanism must be responsible for the formation of Greenland Sea Deep Water. They suggest that the deeper Atlantic water (relatively warm and saline) loses heat by conduction to the Greenland Gyre Water above it (cool and less saline) at a greater rate than it loses salt. The result is that the bottom of the Gyre Water, being heated from below, becomes unstable and convects upward carrying heat to the surface and discharging it to the atmosphere. The Atlantic Water which is cooled by this process at its contact with the Gyre Water, without losing salt, becomes more dense and sinks to form the GSDW. The exchange across the interface between the Atlantic Water and the Gyre Water is identified as a double diffusion process. However, there is some question about this since a typical double diffusion process is a short-range one (decimetres to metres) whereas the proposed Greenland Sea process requires heat transfer over 50 to 200 m. Note that if it is an example of double diffusion, it is different from the one described earlier in that in the present case no mixing occurs between the layers—there are simply two convection cells developing, one in each layer.

7.45 Labrador Sea, Baffin Bay and Hudson Bay

The East Greenland Current (Fig. 7.19) carries much ice from the Arctic down the coast of Greenland, maintaining the low temperatures and rendering access to the east coast of Greenland difficult. The Current flows round the southern tip of Greenland into the Labrador Sea, having picked up some Atlantic water south-west of Iceland. It continues north up the west coast as the *West Greenland Current* from which water branches off to the west until the Current eventually peters out in Baffin Bay. The inflow to this area is balanced by the southward flow, along the west side of Baffin Bay, of the *Baffin Land Current* which continues south as the *Labrador Current* down the west side of the Labrador Sea back into the Atlantic. In this region the differences between the properties of the in- and outflowing currents are not as great as to the east of Greenland. The West Greenland Current has temperatures around $2°$ C and salinities of 31 to $34°/_{oo}$, while the Labrador Current water is at $0°$ C or less and 30 to $34°/_{oo}$. The frequently quoted calculations of Smith, Soule and Mosby (1937) for the Labrador Sea indicate that above 1500 m depth the total inflow is 7.5 Sv, comprised of the East Greenland Current from the south (5 Sv), the Baffin Land Current from the north (2 Sv) and the inflow from Hudson Bay (0.5 Sv). The outflow totalling 5.6 Sv consists of 1 Sv northward along the west Greenland coast to Baffin Bay and 4.6 Sv south in the Labrador Current. They concluded that the balance of 1.9 Sv must flow out as deep water from the Labrador Sea into the Atlantic. Other estimates have indicated that considerable variations in the net flow may occur but the concensus is that in the long-term average there is a net inflow in the upper layers and an outflow of deep water.

The Labrador Sea is another region where it has been assumed that deep, convective overturn as a result of winter cooling was the mechanism for deep-water formation. As some winter cruises did not show the vertically homogeneous water mass which would be expected in this process, a winter cruise in C.S.S. *Hudson* in 1966 (Lazier, 1973) was designed to study the region with closely spaced stations in case the formation area was very small in horizontal extent. No evidence of large-scale convective overturn was obtained and a more plausible explanation of the data was that surface cooling resulted in a flow downward along moderately sloping σ_θ surfaces rather than vertically. However, in 1967 data from Weather Station BRAVO in the centre of the Labrador Sea did show evidence of deep convective overturn to 1500 m and it is possible that both vertical overturn and flow along σ_θ surfaces may at different times play their part in the formation of deep water there.

Some contribution to the Labrador Current comes from Hudson Bay (Fig. 7.19). This is an extensive body of water averaging only about 90 m in depth with maximum values of about 200 m. The Bay is usually covered with ice in the winter but free for a time in summer. As a result, most of the oceanographic data is for the summer, mainly from cruises in 1930, 1960 and 1961. There is considerable seasonal river runoff into the Bay from the south and east sides, giving rise to a marked horizontal stratification and an estuarine-type circulation. In summer, the upper water properties range from $1°$ to $9°$C and 25 to $32°/_{oo}$ while the deeper water is from -1.6 to $0°$C and 32 to $33.4°/_{oo}$. The low salinities are generally in the south and east, near the main sources of runoff and consistent with a general anticlockwise circulation in the upper layer. A few observations taken in winter through the ice indicate upper salinities from $28°/_{oo}$ in the south-east to $33°/_{oo}$ in the north, with temperatures everywhere at the freezing point appropriate to the salinity. The implication is that the waters are vertically mixed each year; the high dissolved oxygen values of 4.5 to 8 mL/L in the deep water are consistent with this.

The Labrador Sea is open to the Atlantic to depths of over 4000 m but Baffin Bay, with a maxium depth of 2200 m, is separated from the Labrador Sea, and hence from the Atlantic, by the sill in the Davis Strait of depth only about 600 m. The deep waters in Baffin Bay below sill depth are markedly different from those at the same depths outside. The temperature is between $0°$ and $-0.5°$C and the salinity $34.5°/_{oo}$ with relatively low oxygen values of 4 mL/L. Sverdrup interpreted the *Baffin Deep Water* as Labrador Sea water which had moved north below the surface, mixed with cool, low salinity surface water, and then had its density increased by freezing to sink to fill the basin. Since Baffin Bay is ice-covered during the winter there is little information for this season but there are indications that water having the properties of the Deep Baffin Water does not occur at the surface. An alternative suggestion by Bailey (1957) is that the source of the Baffin Deep Water is probably inflow from the Arctic through Nares Strait, the passage between Greenland and Ellesmere Island. Water of the same salinity and temperature as the Baffin Deep Water certainly occurs at the appropriate depth in the Arctic Sea. The annual inflow to Baffin Bay is presumably relatively small so that the water in the basin has a long residence time there and the oxygen content gets depleted.

7.46 Adjacent seas; inflow and outflow characteristics

Sverdrup contrasted the horizontally separated inflow and outflow of the northern seas with the vertically separated flows into and out of the

Mediterranean but, as will be demonstrated below, the distinction is not clear cut.

Figure 7.20 shows schematically some of the flow characteristics for seas or other bodies of water adjacent to a main ocean. Three examples of seas with vertically separated exchange (type A) are illustrated. The European Mediterranean, and the Red Sea to be described later, have upper layer inflow and deeper outflow over their sills. In both cases, the excess of evaporation over fresh-water input increases the salinity of the upper waters. From the Mediterranean, the outflow is of the Levantine Intermediate Water from

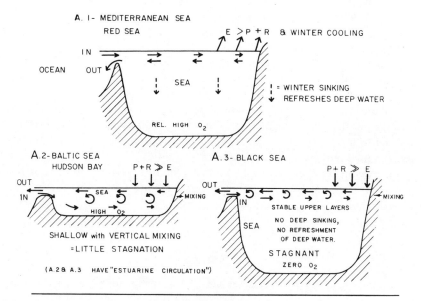

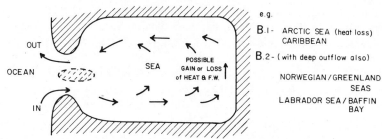

FIG. 7.20. *Schematic diagram of various types of circulation in seas adjacent to oceans.*

depths of a few hundred metres only, the deeper water being stratified. In the Red Sea, the deep waters are much less stratified but the outflow is still from a shallow intermediate layer. The Baltic Sea provides an example of estuarine circulation (Chapter 8) with surface outflow driven by an excess of fresh-water input over loss by evaporation, but with significant vertical mixing in the basin. Most fjords follow this pattern. The Black Sea is an extreme example of estuarine circulation in which the upper layer is of low density and high stability to below sill depth, so that little vertical circulation takes place. A few fjords also possess these properties.

A clear example of the second type (B, Fig. 7.20) is the American Mediterranean (the Caribbean Sea) into which inflow is through the Antilles Islands in the south-east and outflow from the north-west through the Straits of Florida. Less strict examples of type B are the northern seas adjacent to the North Atlantic. For the Norwegian/Greenland Seas, the water of the North Atlantic Current which becomes the Norwegian Current has temperatures from 4° to 13° C and salinities from 34.9 to 35.3‰, whereas the outflowing East Greenland Current water is at $-1.5°$ to 2° C and 31 to 34‰. On a T–S diagram the upper waters of the Norwegian Current are represented by a line roughly parallel to the temperature axis, with the density mainly determined by temperature, whereas the upper-layer water of the outflowing East Greenland Current is represented by a line roughly parallel to the salinity axis with the density determined almost entirely by salinity as is characteristic of high latitude waters. Another example of a type B basin is the Labrador Sea/Baffin Bay region in the north-west.

The reason why these cannot be considered as strict type B (horizontally separated) flows is that in both cases there are now known to be large deep-water outflows as well as the horizontally separated upper-layer flows.

7.5 Arctic Sea

Our knowledge of the Arctic Sea has developed considerably since the mid-1950s. Numerous soundings and oceanographic stations have been taken from ships, as well as through the ice from semi-permanent camps on ice islands or ice floes or from temporary camps established by aircraft transportation. In particular it has been demonstrated that the Arctic Sea is divided into two basins by the Lomonosov Ridge which extends from Greenland past the North Pole to Siberia (Fig. 7.19). These two basins have been named the Canadian Basin (depth about 3800 m) and the Eurasian Basin (depth about 4200 m). Soundings along the Lomonosov Ridge are not numerous enough to determine its sill depth with any certainty but comparisons of water properties on either side suggest that it is at 1200 to 1400 m below sea level. A feature of the bottom topography is the broad continental shelf off Siberia, 200 to 800 km wide and occupying about 36 % of the area of the Arctic Sea but containing only 2 % of the total volume of water in the Sea. The main connection with the other oceans is with the Atlantic through the gap between Greenland and Spitsbergen with a main sill depth there of 2600 m while the sill depth between Spitsbergen, Franz Josef Land and Novaya Zemlaya is only about 200 m (Coachman and Aagaard, 1974). The Bering Strait connection to the Bering Sea and the Pacific Ocean has a sill depth of about 45 m and is narrow, but the water flow into the Arctic is not insignificant. There are also connections from the Arctic through the Canadian Archipelago by several channels, principally Nares Strait (sill depth 250 m) and Lancaster Sound (sill depth 130 m) which lead to Baffin Bay and thence to the Atlantic. These passages are difficult of access because of ice and are not fully charted.

7.51 Arctic Sea: upper-layer circulation

Information on the circulation of the upper layers has been obtained from the records of the movements of ships held in the ice, such as the *Fram* and the *Sedov*, and from movements of camps on the ice. In addition, geostrophic calculations have been made from the water-density distribution. These various sources yield a consistent picture of the surface-layer movement (Fig. 7.19) which is best described as a clockwise circulation in the Canadian Basin leading out to the East Greenland Current and, in the Eurasian Basin, a movement by the most direct path toward Greenland and out in the East Greenland Current. The speeds are of the order of 1 to 4 cm/s or, perhaps, more meaningfully stated as 300 to 1200 km/year when considered in relation to the size of the Arctic Sea which is about 4000 km across. The speed and distance may be compared to the 3 years taken by the *Fram* to drift from north of the Bering Strait to Spitsbergen, and the $2\frac{1}{2}$ years for the *Sedov* to drift about 3000 km. The movement is not by any means steady but has frequent variations of speed and direction which average out to the figures quoted.

7.52 Arctic Sea water masses

Three main water masses are recognized (Coachman and Aagaard, 1974). These are (Table 7.1 and Fig. 7.21 (a)) the surface or *Arctic Water* from the sea surface to 200 m depth, the *Atlantic Water* from 200 to 900 m, and the *Bottom Water* from 900 m to the bottom. One of the features of the water structure is that its density is determined largely by the salinity (which is why the marked temperature maximum of the Atlantic Water can exist).

TABLE 7.1

Arctic Sea water masses

Water mass		Properties	
Name (circulation direction)	Boundary depth		Seasonal variation
	Surface		
ARCTIC SURFACE		T: Close to F.P., i.e. − 1.5 to − 1.9° C S: 28 to 33.5°/₀₀	ΔT: 0.1 K ΔS: 2 °/₀₀
	25 to 50 m		
ARCTIC SUBSURFACE		T: Canadian Basin − 1 to − 1.5° C Eurasian Basin − 1.6° to 100 m, then increase S: Both basins 31.5 to 34°/₀₀	Small
	100 to 150 m		
ARCTIC LOWER		Intermediate between Subsurface and Atlantic	
	200 m		
(all above masses circulate clockwise)			
ATLANTIC (anticlockwise)		T: Above 0° C (to 3° C) S: 34.85 to 35°/₀₀	Negligible
	900 m		
BOTTOM (uncertain, small)			2000 m Bottom
		T: Canadian Basin Eurasian Basin S: Both Basins	− 0.4° − 0.2° C (rise adiabatic) − 0.8° − 0.6° C 34.90 to 34.99°/₀₀
	Bottom		

7.521 Arctic Water

The Arctic Water (0 to 200 m) can be divided into three layers which will be called the *Surface Arctic*, the *Sub-surface Arctic* and the *Lower Arctic Waters*.

Surface Water (Table 7.1) is much the same across the whole Arctic and extends from the surface to between 25 and 50 m depth. The salinity is strongly influenced by the freezing or melting of ice and has a wide range from 28 to 33.5°/₀₀. The temperature is also controlled by the melting or freezing which involves considerable heat transfer at constant temperature (the freezing point). In consequence the temperature remains close to the freezing point of sea-water which varies only from − 1.5°C at a salinity of 28°/₀₀ to − 1.8°C at a salinity of 33.5°/₀₀. Seasonal variations in water properties are limited to this layer and range up to 2‰ in salinity and 0.2K in temperature.

The *Subsurface* layer (25/50 m to 100 m) in the Eurasian Basin is isothermal

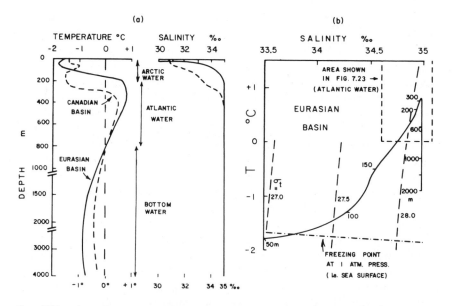

FIG. 7.21. Arctic Sea: (a) typical temperature and salinity profiles for the two basins, (b) T–S diagram for the Eurasian Basin water.

to 100 m (Fig. 7.22(b) and then increases but there is a strong halocline between 25 and 100 m (Fig. 7.22 (a)). Below 100 m the temperature increases markedly but the salinity only increases slowly. The fact that the Subsurface water is isothermal (limited to the freezing point) but not isohaline indicates that its structure cannot be due to vertical mixing between the Surface layer

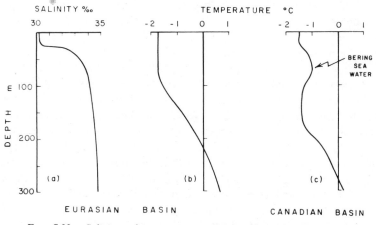

FIG. 7.22. Salinity and temperature profiles for the Arctic Water Mass.

and the deeper layers. It is probable that the Subsurface Water is maintained by horizontal advection (flow) from the Eurasian Shelf. The mechanism suggested by Coachman is that the saline Atlantic Water which enters near Spitsbergen continues below the surface along the Eurasian continental slope which is indented by several deep submarine canyons. At the same time the considerable river runoff from northern Asia flows north over the shelf as a cold, low-salinity surface layer. It mixes at its subsurface contact with the warmer, more saline Atlantic Water to form the Subsurface Water which is close to its freezing point. The Subsurface Water continues out into the Arctic Sea to maintain the layer there between 25 and 100 m. The canyons are necessary to feed the saline Atlantic Water into the shelf area, and the vertical mixing process is similar to that which occurs in an estuary where fresh river water flows over saline sea-water as described in Chapter 8.

The Subsurface Water in the Canadian Basin also shows a halocline from 25 m to 100 m but its temperature structure is different from that in the Eurasian Basin. There is a characteristic temperature maximum at 75 to 100 m depth (Fig. 7.22 (c)) with a consequent temperature minimum of $-1.5°C$ at about 150 m and then an increase to the deeper water values. The temperature maximum is attributed to Bering Sea water coming into the Arctic through the Bering Strait. This water is warmer than the Arctic Surface layer but slightly denser because of its salinity. It presents one of the relatively few examples of a subsurface temperature maximum occurring in the open ocean. The reason that it occurs here is because the water is close to its freezing point and the effect of salinity preponderates over that of temperature in determining density. The temperature maximum is found to be most prominent in the Chukchi Sea north of the Bering Strait, and it diminishes around the clockwise circulation of the Canadian Basin.

The *Arctic Lower Water* is essentially a mixing layer with properties intermediate between the Subsurface Arctic Water above and the Atlantic Water below.

7.522 Atlantic Water

The second water mass, the *Atlantic Water* (Fig. 7.21 (a) vertical profiles, (b) T–S diagram), is recognized chiefly by having a higher temperature than the water above or below it. Where it enters as the *West Spitsbergen Current*, on the east side of the Greenland–Spitsbergen gap, its temperature is up to $3°C$ and its salinity 34.8 to $35.1°/_{oo}$. In the Arctic Sea its temperature decreases gradually to $0.4°C$ and its salinity is within the limiting range from 34.85 to $35°/_{oo}$. Its movement has been traced by the core method along the Eurasian continental slope, with some water branching off to the north and out as part of the East Greenland Current. The remainder flows across the Lomonosov Ridge into the Canadian Basin.

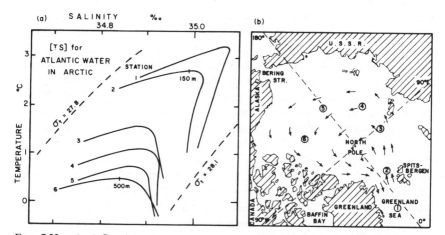

FIG. 7.23.   *Arctic Sea: (a) temperature maximum part of T–S diagram for core method analysis of flow direction for Atlantic Water, (b) circulation inferred from successive erosion of core shown in (a) for stations 1 to 6.*

The application of the core method is as follows. Figure 7.21 (b) shows a T–S diagram for the water column in the Eurasian Basin with the marked temperature maximum of the Atlantic Water near the salinity of 35°/$_{oo}$. (Note that the temperature maximum is emphasized by the magnified temperature scale in this diagram, 2K equivalent on paper to 1‰ salinity, compared with the other T–S diagrams shown previously, e.g. Fig. 7.14 where 10C° was equivalent to 1°/$_{oo}$.) Then Fig. 7.23 (a) shows the temperature maximum part of the T–S diagram (outlined in Fig. 7.21 (b)), the core of the Atlantic Water, for a selection of stations from the Greenland Sea and around the Arctic basins (Fig. 7.23(b)) to show how the core is gradually eroded away by mixing during its circuit. The water mass itself appears to remain in the depth range from 200 to 900 m, but the depth of the temperature maximum increases from about 150 m at the top of the Atlantic Water just before entry into the Arctic near Spitsbergen to nearly 500 m in the Canadian Basin. The reason is that the temperature gradient in the upward direction is greater than in the downward direction (Fig. 7.21 (a)) with the result that more heat is lost upward from the layer than downward and the temperature maximum increases in depth. (There may also be some descent of the Atlantic Water.) The circulation of the Atlantic Water is then basically counter-clockwise around the Arctic Sea, the opposite direction to that of the Arctic Water above it.

7.523 Bottom Water

The Arctic Bottom Water (Fig. 7.21; Table 7.1) extends from the lower 0°C isotherm, about 900 m depth, to the bottom and comprises about 60 % of the total water volume of the Arctic Sea. Its salinity range through the whole volume is very small, from 34.90 to $34.99°/_{oo}$, and in any particular area the change in the vertical direction is generally smaller than this. There is a tendency for the salinity to increase very slightly with depth. The *in situ* temperature varies over a range of 0.2K in the vertical column. In the Eurasian Basin the temperature reaches a minimum of $-0.8°C$ at 2500 m, while in the Canadian Basin the minimum is $-0.4°C$ at 2000 m (Fig. 7.21(a)). Below the minimum the temperature rises by about 0.2K to the bottom. The rate of increase is equal to the adiabatic rate, i.e. it can be attributed entirely to the compression of the water as it sinks. It is believed that this Bottom Water originates in the Norwegian Sea and flows thence into the Arctic. The reason for the higher temperature of the water in the Canadian Basin than in the Eurasian Basin is that the former is water which has come across the Lomonosov Ridge (Fig. 7.19) at depths not greater than 1200 to 1400 m. The deeper, colder water which gets into the Eurasian Basin is prevented by the Lomonosov Ridge from entering the Canadian Basin.

7.53 Arctic Sea budgets

The Arctic Sea has always attracted the attention of oceanographers wishing to exercise their talents by investigating the water budget, and many sets of calculations have been made. The results obtained vary somewhat in detail but the main features are substantially the same. A recent calculation by Aagaard and Greisman (1975) has been chosen for demonstration because it also includes heat and salt budgets (Table 7.2). This does not represent an absolute budget in the sense that all the flow terms and water properties are known exactly; in fact, information on most of the quantities is sparse and, in any case, variations with time are likely to occur. Nevertheless, it is worth making even trial budgets in order to determine which terms appear to be important and where one's observational efforts should be directed in order to improve the budgeting.

From the budget in Table 7.2, it can be seen clearly that the main volume fluxes are through the Greenland–Spitsbergen gap but that the Bering Strait and Arctic Archipelago flows are not negligible. The main heat flux (larger than in previous budgets) is carried by the West Spitsbergen Current and overall there is a net advective inflow (Q_v positive) to the Arctic; this must be lost through the sea surface. (Note that because of its high latent heat of fusion an outflow of ice is equivalent to an inflow of heat and vice versa.)

Calculations of heat budgets are done for other reasons than for displaying virtuosity in deduction and calculation. From the water budget, in relation to the volume of the basin itself, one can obtain an idea of the rate at which the water in the basin is exchanged. This may be important for determining the rate of replenishment of nutrients in an area important for fisheries or for the

TABLE 7.2

Annual mean water, heat and salt budgets for the Arctic Ocean (Adapted from Aagard and Greisman, 1975)

	Mean water properties		Transports		
	Temp. (°C)	Sal. (°/oo)	Volume (Sv)	Heat 10^{12} W)	Salt (10^6 kg)
Bering Strait, Water	0.5	32.4	1.5	4	49
Ice	−10	3	negl.	−2	negl.
Arctic Archipelago	−0.7	34.2	−2.1	5	−72
East Greenland Current					
Polar Water	−1.2*	34.0	−1.8	8	−61*
Atlantic Water	0.5*	34.9	−5.3	−13	−185*
Ice	−10	3	−0.1	34	
West Spitsbergen Current	2.2	35.0	7.1	68	249
Spitsbergen–Fr. Josef Land	2.7	34.9	−0.1	−1	−4
Fr. Josef Land–Nov. Zemlaya	0.9*	34.7	0.7	3	24
Runoff	5.0	0	0.1	2	0
Total Inflow/heat gain	1.8	34.6	9.4	124	322
Total outflow/heat loss	−0.1	34.6	−9.4	−16	−322
Net exchange			0	108	0

Notes: (1) Positive values represent inflow or heat gain, negative values represent outflow or heat loss.
(2) Heat transport is calculated relative to 0.1°C, the mean outflow temperature.
(3) Items marked with an asterisk (*) have been adjusted, within observed data limits, to satisfy continuity.

rate of removal of sewage or industrial effluent. For the Arctic it is estimated that the surface water is substantially all replaced in a period (residence time) of 3 to 10 years, the deep water in 20 to 25 years (from the above budget) and the bottom water in about 150 years.

7.54 Ice in the sea

Ice in the sea has two origins, the freezing of sea-water and the breaking of ice from glaciers. The majority of the ice comes from the first of these sources and will be referred to as *sea-ice*; the glaciers supply *icebergs*.

7.541 Properties of sea-ice

When sea-water freezes, needle-like crystals of pure ice form first, thereby increasing the salinity of the remaining liquid. However, as the crystals grow they tend to trap some of the concentrated salt solution ("brine") in pockets among them. The result is that new sea-ice in bulk is not pure H_2O but has a salinity of 5 to 15°/oo. The faster the ice forms, the more saline is it likely to be.

If this ice is lifted above sea-level, as happens when ice becomes thicker or when rafting occurs, the brine gradually trickles down through it and eventually leaves almost saltless, clear old ice. Such ice, a year or more old, may be melted and used for drinking whereas new ice is not potable. Sea-ice must therefore be considered to be a material of variable composition and properties, which depend very much on its previous history. (For more detail see Doronin and Kheisin, 1975.)

The freezing point of sea-water decreases from 0°C at a salinity of $0°/_{oo}$ (fresh water) to $-1.91°$C at $35°/_{oo}$ (Fig. 3.1). An associated property is the temperature of maximum density which, at the surface (i.e. atmospheric pressure), decreases from 3.94°C at a salinity of $0°/_{oo}$ to $-1.33°$C at a salinity of $24.7°/_{oo}$. At this temperature, the freezing point coincides with the temperature of maximum density for sea-water; for higher salinities the maximum density occurs at the freezing point. The freezing point decreases with increase of pressure (by about $0.08°$ C per 100 m increase of depth).

The density of pure water at 0°C is 999.9 kg/m³ and that of pure ice is 916.8 kg/m³. However, the density of sea-ice may be greater than this last figure (if brine is trapped among the ice crystals) or less (if the brine has escaped and gas bubbles are present). Values from 924 to 857 kg/m³ were recorded on the Norwegian *Maud* Expedition (Malmgren , 1927).

Malmgren gives values for the specific heat of sea-ice of as much as 67 kJ/(kg C°) at $-2°$C and $15°/_{oo}$ salinity. This surprisingly high figure arises because the measurement of the specific heat requires heating the sample over a finite range of temperature and there will be some melting of ice crystals into the brine. The 67 kJ/(kg C°) then includes some latent heat of melting and the high values are not true specific heats in the sense that the term is used for pure substances. The latent heat of melting decreases from 335 kJ/kg (80 cal/g) at 0°C and $0°/_{oo}$ salinity to only 63 kJ/kg at $-1°$C and $15°/_{oo}$ salinity.

In the sea, incipient freezing is indicated by a "greasy" appearance of the sea surface due to the presence of flat ice crystals. As freezing continues, individual plates or spicules of ice develop in quantity (*frazil ice*) and these tend to aggregate to form *slush ice*. This slush then further aggregates to form *pancake ice*, flat rounded sheets of ½ m or more in diameter. These then freeze together to form *floe* or *sheet ice*. Clear sheet ice such as forms on fresh-water ponds does not form on the sea. The rate of formation of ice depends very much on conditions. It is favoured by low salinity, lack of mixing by wind or currents, shallow water, and the presence of old ice which keeps the water calm. At a temperature of $-30°$C, 3 cm of ice can form in 1 hour and 30 cm in 3 days, the rate of increase of thickness diminishing as the ice thickness increases because ice is a poor conductor of heat. In Canadian Arctic waters a total of 2 to 3 m of ice may form inshore between September and May and then all melt by July.

7.542 Distribution of sea-ice

In the Arctic region the sea-ice may be divided into three categories. The most extensive is the *Polar Cap Ice* which is always present and covers about 70 % of the Arctic Sea, extending from the pole approximately to the 1000-m isobath. It is very hummocky and, on the average, several years old. Some of this cap ice melts in the summer and the average thickness decreases to about $2\frac{1}{2}$ m. Open water spaces, *polynyas*, may form. In the autumn these freeze over and the ice in them gets squeezed into ridges, or pushed so that one piece slides up over another (*rafting*). In the winter the average ice thickness is 3 to $3\frac{1}{2}$ m but hummocks increase the height locally up to 10 m above sea level with increase in depth below sea level of four to five times this. The hummocks build up as the result of inward forces on ice sheets causing them to buckle or raft. The occasional ice islands which have fairly uniform thickness considerably greater than the regular cap ice originate from glaciers in northern Ellesmere Island.

Although this polar cap ice is always present it is not always the same ice. Up to one-third of the total cap and pack ice is carried away in the East Greenland Current each year while other ice is added from the pack ice described below. The polar cap ice circulates in a clockwise direction about a centre somewhere between the pole and northern Canada. The motion is not smooth, but irregular or zig-zag because the cap ice does not move as a solid mass but, under the influence of changing wind stress, its velocity varies and it shears. The polar cap is only penetrable by the heaviest ice-breakers.

The *Pack Ice* lies outside the polar cap and covers about 25 % of the Arctic area. It is lighter than the cap ice and its area varies somewhat from year to year but extends usually to about the 1000-m isobath or further inshore. It penetrates further south in the East Greenland and Labrador Currents by which it is carried. Its areal extent is least in September and greatest in May. Some of it melts in summer and some gets added to the cap by rafting. The pack ice can be penetrated by ice-breakers.

Lastly, the *Fast Ice* is that which forms from the shore out to the pack. This ice is "fast" or anchored to the shore and extends out to about the 20-m isobath. In the winter it develops to a thickness of 1 to 2 m but breaks up and melts completely in summer. When it breaks away from the shore it may have beach material frozen into it and this may be carried some distance before being dropped as the ice melts giving rise to "erratic" material in the bottom deposits.

The ice which prevents or impedes navigation in the northern parts of the Canadian Archipelago, along the east coast of Greenland, in Baffin Bay, the Labrador Sea area and in the Bering Sea is the pack ice. It is of only a few metres thickness; separate pieces are called *floes* and should not be referred to as icebergs.

In the North Pacific itself, ice does not occur, but it is formed in the adjacent seas to the north and west, the Bering Sea, the Sea of Okhotsk and the north of the Sea of Japan. In the Bering Sea pack ice extends in winter to about $58°$ N but clears completely in the summer, retreating through the Bering Strait to about $70°$ N

7.543 Icebergs

The icebergs which are a menace to ships in the North Atlantic and in the Antarctic originate as pieces "calving" (i.e. breaking) off glaciers. In the North Atlantic the majority of the icebergs come from the glaciers of the west coast of Greenland and the east coast of Ellesmere Island. Icebergs differ from pack ice both in their origin on land and in their much greater vertical dimensions. When calved off the Greenland glaciers they frequently extend to 70 m above sea level but this height decreases rapidly thereafter. They are eventually carried south in the Labrador Current and some pass into the North Atlantic off Newfoundland. Here they are usually a few tens of metres high, the highest recorded being about 80 m high and the longest about 500 m. The density of their ice is about 900 kg/m^3, a little less than that of pure ice because of the gas bubble content. The ratio of volume below sea level to that above is close to 7 to 1 but the ratio of maximum depth below sea level to height above it is less than this depending on the shape of the iceberg. It varies from 5 to 1 for a blocky shaped berg to only 1 to 1 for very irregular bergs. North Atlantic icebergs are generally referred to as *pinnacled bergs*.

The drift of icebergs is generally determined chiefly by the water currents, whereas the shallower pack ice is more directly influenced by the wind stress. This does not mean that the pack ice moves in the direction of the wind. In the northern hemisphere it moves significantly to the right of the wind direction due to the effect of the Coriolis force. The fact of the movement to the right of the wind direction was known to sailors long before it was explained scientifically by Nansen in a qualitative manner and, at his suggestion, by Ekman quantitatively (see Section 8.2 for a simple description and texts in dynamical oceanography for more detail).

In the Southern Ocean, the immediately available information on the distribution of ice is not great, although in recent years a considerable amount of material has been gathered using satellite observations in conjunction with ship reports. On the basis of available data it appears that icebergs may be found to between $50°$ and $40°$ S while pack ice extends only to $65°$ to $60°$ S. The relatively zonal distribution is probably due to the zonal character of the currents of the Southern Ocean. An outstanding feature of the Antarctic is the Ross Ice Barrier with a sea front of about 700 km to the Pacific and a height of 35 to 90 m above sea level with a corresponding depth below. This *shelf ice* represents the extension of the glaciers on the Antarctic continent out on to the sea where they float until bergs break off. These *tabular bergs* may be up to 80 to 100 km long and tens of kilometres wide.

7.6 Pacific Ocean

Our oceanographic knowledge of the Pacific Ocean is less complete in some respects than that of the Atlantic. The North Pacific can be described well, and our information on the disposition of currents in the Equatorial Pacific appears to be good but quantitative information on their volume transports is very limited. Data on the South Pacific are sparse and it is not possible to describe it as adequately as the South Atlantic.

7.61 Pacific Ocean circulation

The circulation of the upper waters of the Pacific (Fig. 7.24) as a whole is very similar in its main features to that of the Atlantic. There is a clockwise gyre in the North Pacific and a counterclockwise one in the South Pacific, with an equatorial current system between them. The whole equatorial current system is well and clearly developed in the Pacific and it will be described first, and details of the two gyres given later.

7.611 Pacific Ocean: equatorial circulation

The Pacific equatorial current system (Fig. 7.24) is now recognized to include at least four major currents, three of which extend to the surface and one of which is below the surface. The three major upper-layer currents (evident at the surface) are the westward-flowing *North Equatorial Current* (NEC) between about 20° to 8°N (in mid-Pacific at 160°W), the westward *South Equatorial Current* (SEC) from about 3°N to 10°S and the narrower *North Equatorial Countercurrent* (NECC) flowing to the east between them. The fourth current, the *Equatorial Undercurrent* (EUC), flows to the east below the surface straddling the equator from about 2°N to 2°S. This equatorial system can be traced from the Gulf of Panama in the east almost to the Philippines in the west, a distance of some 15,000 km.

The surface current system is driven by the trade winds and is asymmetrical about the equator because the trade-wind system is asymmetrical as is shown in Fig. 7.25, a situation which is related to the asymmetrical distribution of land and sea about the equator. Although the trade wind system is present throughout the year it undergoes some seasonal variations which will be described. (In doing so we will identify the seasons by month names when considering areas which include both hemispheres to avoid the ambiguity of the terms "winter" and "summer".) In February, the monsoon season, the north-east trade winds stop short of the equator at about 5°N in

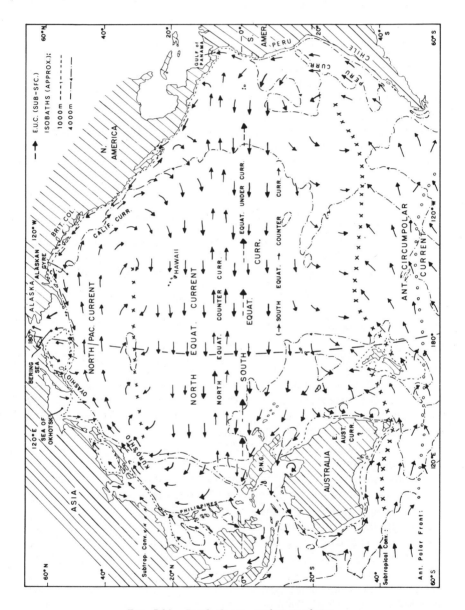

FIG. 7.24. *Pacific Ocean—surface circulation.*

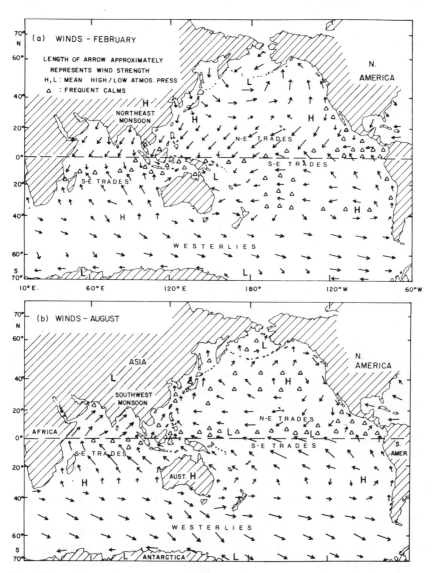

FIG. 7.25. *Indian and Pacific Oceans—winds and mean atmospheric pressure highs and lows: (a) February (north–east monsoon, (b) August (south–west monsoon).*

the east but extend to the equator in the west. At this time, the south-east Trades extend across the equator in the east. In August, the southern limit of the north-east Trades is at 10°N while the south-east Trades cross the equator to about 5°N across the whole Pacific. (The seasonal change of the equatorial wind system is more marked in the Indian Ocean and will be discussed later.) It should be noted that the region between the trade winds, the "doldrums", is a region of reduced winds with 10 to 30% calms, not a region of continual calm.

To the south of the Hawaiian Islands, the volume transport of the NEC is about 45 Sv and the surface speeds are of the order of 25 to 30 cm/s (20 to 25 km/day). The SEC is well known as a surface current to the west with speeds of the order of 50 to 65 cm/s. Recent repeated measurements over a period of 15 months during the Tahiti Shuttle Experiment between Tahiti and Hawaii (Firing, 1981; see Wyrtki et al., 1981, for a description of the Experiment) indicated an average mean transport of 17 Sv with an annual cycle of amplitude ± 10 Sv about the mean, with speeds higher at 2° N that at the equator. Both of these currents show some seasonal variation of speed and direction but are consistent in their direction to the west. The eastward-flowing NECC is more variable than these. The surface speed is from 35 to 60 cm/s except in March and April when it decreases to 20 cm/s or less. In 1958 measurements (Knauss, 1961) at 107° W showed the NECC to extend through the thermocline to depths of some 1500 m. If this pattern were characteristic of the full width of the current it would indicate a total transport to the east of 60 Sv, twice the previously accepted figure of 30 Sv for the upper layers only. However, in 1959 at about 120° W, the small eastward transport in the upper 125 m was balanced by a westward transport between this depth and 1500 m with a resultant transport of only 1 Sv, a remarkable change from the previous year. During the Tahiti Shuttle Experiment (1979–80), the transport varied from 5 to 35 Sv with typical surface speeds of 45 cm/s (Wyrtki, 1980).

These three currents complete the classical picture of the equatorial current systems, but in recent years we have become aware of further components. In 1952, the fourth major equatorial current, the Equatorial Undercurrent (EUC) or Cromwell Current, was recognized flowing to the east at the equator below the surface and embedded in the SEC. Weaker and less well-defined components are the South Equatorial Countercurrent (SECC) to the east, described first by Reid (1959), from roughly 9° to 11° S (in mid-Pacific), and a weak westward current from about 11° to 16° S.

The last two are weak currents which have been located as geostrophic flows. For the SECC, speeds of 15 to 30 cm/s and a transport of about 10 Sv have been estimated at 170° E but the current appears to be much weaker in the eastern Pacific. Because of the paucity of direct observations in the eastern South Pacific these currents do not appear on atlases of sea-surface currents. In addition (not shown in Fig. 7.24) there are indications in the western Pacific of an eastward North Tropical Countercurrent north of the NEC, and possibly of an eastward South Tropical Countercurrent to the south of the SEC (Donguy and Henin, 1975).

A recent review of the equatorial current systems has been presented by Leetma, McCreary and Moore (1981).

7.612 Pacific Ocean: Equatorial Undercurrent

The Equatorial Undercurrent (Fig. 7.24) is almost as large in volume transport as the westward-flowing equatorial currents and it lies only 100 m or less below the sea surface along the equator, yet it was not discovered until 1952 (Cromwell, Montgomery and Stroup, 1954; Montgomery, 1962). The programme of the Pacific Oceanic Fisheries Investigation of the U.S. Fish and Wildlife Service included systematic studies of the equatorial regions. In the summer of 1952 a cruise planned for detailed study at 150° W included current measurements by free drifting drogues at and below the surface at 3° S, at the equator and at 7° N. The outstanding result from the drogue measurements was that near the equator and embedded in the well-known westward-flowing SEC there was a previously undiscovered eastward-flowing current between 70 and 200 m depth. The first intensive study was made in 1958 by Knauss (1960), then with the Scripps Institution of Oceanography, along the equator between 140° W and 89° W. Subsequently, reviews of earlier data for the western Pacific and direct observations, particularly by Rotschi (1970, 1973) and his colleagues from New Caledonia, have shown that the current is continuous from at least 142° E (north of Papua New Guinea) and may start further west. This would mean that it has a length of at least 14,000 km. The current is like a thin ribbon, being only 0.2 km thick but 300 km wide, extending from about 1.5° N to 1.5° S. Even in the oceans where features are typically thin, i.e. small in vertical dimensions relative to horizontal, a current with a thickness to width ratio of 1 to 1500 is remarkable, particularly when it can be followed for 14,000 km. (The ratio is about the same as the thickness to length ratio of the longer side of this page.) Speeds of up to 170 cm/s have been measured at the current core which rises from about 200 m depth in the west to 40 m in the east near the Galapagos Islands (Fig. 7.26(a)) and may even break surface there at times. (Note that Fig. 7.26(a) does not show the current speed as such but is a plot of the pycnocline, in terms of thermosteric anamaly ($\Delta_{S,T}$), with which the EUC core is closely associated.) Measurements over 15 months during the Tahiti Shuttle Experiment (Wyrtki, 1980) gave a maximum speed of 150 cm/s and a maximum volume transport of 70 Sv, with 60 Sv for a period of 3 months. The annual average was about 40 Sv with an annual variation of ± 13 Sv, the maximum occurring in July and the minimum in January. The NECC and the SEC were 180° out-of-phase with the EUC. It should be pointed out that dynamical studies of the EUC have explained its role as a compensation flow for some of the water transported west by the NEC and SEC.

In the western Pacific (west of 174° E) the current system near the equator (Fig. 7.26(b)) is fairly complicated and variable, the latter feature being related to monsoon wind variations. During January to March when the south-east Trades blow (to the west) there is generally at the surface the westward flow of the SEC. Below this is the eastward-flowing EUC which often has two cores.

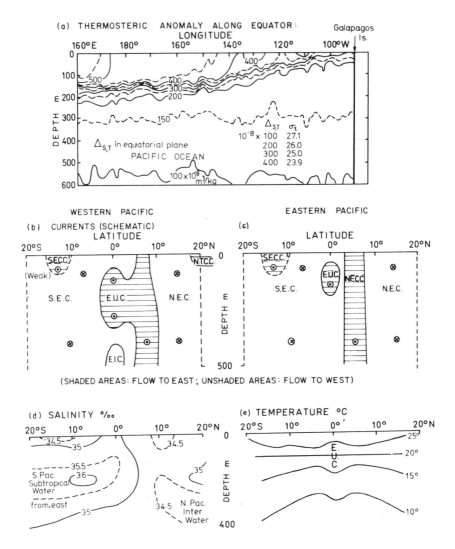

FIG. 7.26. *Equatorial Pacific Ocean: (a) west–east section along equator of thermosteric anomaly, (b, c) south–north sections across currents in western and eastern Pacific (schematic), (d) south–north section for salinity in west, (e) south–north section for temperature in east (schematic).*

The upper of these is at the bottom of the mixed layer and the lower is in the thermocline (which is also the pycnocline). Below this is an Equatorial Intermediate Current (Hisard and Rual, 1970; Hisard, Merle and Voituriez, 1970) to the west and below this again is a deep eastward flow (unnamed as yet). During the monsoon season (January to March) the winds near the equator may even reverse and blow toward the east. Then there can be an eastward flow at the surface, a westward flow below this and then the lower eastward core of the EUC. The Intermediate and deep currents continue to flow. Another characteristic in the west is that the NECC and the EUC are usually joined below the surface (water property differences distinguish them). The vertical section of salinity in Fig. 7.26(d) is included to show that the EUC receives its high-salinity water from a core in the South Pacific.

In the eastern part of the Pacific, although they vary in strength, the southeast Trades continue through the year, and the seasonal reversal of current at the surface does not occur. The pattern is simpler (Fig. 7.26(c)) with a single core EUC flowing to the east embedded at thermocline depth in the westward SEC. The deeper westward Equatorial Intermediate Current is still recognized.

The presence of the EUC is revealed in north–south temperature sections across the equator by a spreading of the isotherms as shown schematically in Fig. 7.26(e) and in actual section in Fig. 7.27(c). In this figure are shown cross-sections at 170° E, taken by Rotschi *et al.* (1972) of (a) the east–west velocity component, (b) salinity, (c) temperature and (d) dissolved oxygen. In this case, the 20° C isotherm is approximately level across the equator through the centre of the EUC core at 200 m depth while higher value isotherms bow upward above the core and lower ones downward below it. Farther east the level isotherm tends to be of higher value. The oxygen isopleths (Fig. 7.24.d) also bow upward and downward but generally the lower ones bow down more than the upper ones bow up. The reason for the difference between the behaviour of the temperature and oxygen isopleths is not known. The salinity structure (Fig. 7.24.b) is quite different from those of the other properties, there being a high salinity core on the south side of the current core. This is Subtropical Water formed farther east near French Polynesia where it sinks and moves westward and equatorward and is then entrained into the Undercurrent.

The EUC is "tied" to the equator by the restoring effect of the Coriolis force which, on an eastward flowing current, tends to bring the current back toward the equator if the current drifts away from it either to north or to south. However, the current is not always perfectly straight. Direct measurements show significant and varying north or south components of velocity which suggest that the current meanders somewhat about its mean track along the equator.

As it approaches the Galapagos in the east, the Undercurrent becomes variable in all respects, and both west and east of the Galapagos the volume transport is only a tenth of that in the mid-Pacific. The current appears to divide around the islands and reform to their east, the high-salinity core on its south side being found again and the current being detected as close as 110 km from Ecuador. The fate of the water which has left the Undercurrent is not clear.

Once the existence of the EUC had been recognized in the Pacific it was

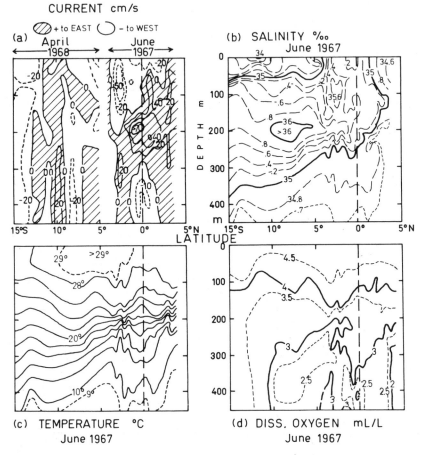

FIG. 7.27. *Western Pacific Ocean—south–north sections at 170° E for: (a) east–west current components, (b) salinity, (c) temperature, (d) dissolved oxygen.*

realized that there has been strong indications of its existence for some time previously. These were in the form of strikingly large wire-angles from the vertical when oceanographic gear was being lowered over the side of ships when near the equator. These wire angles were toward the east and were attributed to the strong, shallow surface current (SEC) carrying the ship toward the west relative to weaker or zero currents below the surface layer. It was only after the Undercurrent had been observed directly with the drogued floats that these observations, together with the observations that at times floating fishing gear drifted toward the east, were recognized as being due to a faster eastward flowing Undercurrent.

The discovery of the Pacific EUC naturally started speculation as to the possibility of there being one in the Atlantic. In 1962 Metcalf *et al.* presented observational evidence for such a current in the eastern equatorial Atlantic. About the same time it was pointed out (Montgomery, 1962) that the Atlantic Equatorial Undercurrent had been clearly observed and recognized as such in 1886 by Buchanan, one of the scientists who had earlier taken part in the *Challenger* expedition. Buchanan's accounts were published in well-known geographical journals but had apparently been forgotten until the independent discovery of the Equatorial Undercurrent in the Pacific stimulated a review of the literature. It has been suggested that remarks criticizing the validity of Buchanan's methods, made in Krümmel's *Handbuch der Ozeanographie* which carried much authority early in the century, may have diverted attention from Buchanan's observations.

Incidentally, the search for an equatorial undercurrent in the Atlantic (and Indian) Ocean was done for other reasons than to permit Atlantic oceanographers to keep up with their Pacific colleagues. It was done more to assist in the search for an adequate dynamic explanation for the existence of an undercurrent at the equator. If an undercurrent were found in all the oceans it would suggest that as it was a typical feature of the equatorial current system then a common explanation might be adequate. Studies of the differences between the equatorial features as well as their similarities would help in deciding between alternative explanations. On the other hand, if the undercurrent were found only in the Pacific it might be that it was a consequence of some peculiarity of that ocean's circulation and this also might help in deciding on the most satisfactory explanation.

Since the equatorial system is an interconnected one, the discovery of the new currents and the new measurements of the volume transports will require a reappraisal of the whole system.

7.613 Convergences and divergences associated with current systems

One other dynamic feature associated with the Coriolis force which affects the distribution of water properties and the plankton, and incidentally of pelagic fishes, in the upper layer is the occurrence of divergences and convergences at the boundary between oppositely directed currents. The reasons are shown schematically in Fig. 7.28. Figure 7.28(a)(1) shows two oppositely directed currents in the northern hemisphere with the more northerly flowing to the west and the other to the east. Then (a)(2) shows that the effect of the Coriolis force, to the right of the direction of motion, is to cause the two currents to *diverge* and consequently subsurface water will rise to replace the diverging surface waters. This occurs between the NEC and the NECC (see Fig. 7.29(b) to be described later). Between two currents flowing as in Fig. 7.28(b)(1), a *convergence* occurs as shown in (b)(2) and as occurs between the NECC and the SEC (Fig. 7.29(b)) with downward movement of

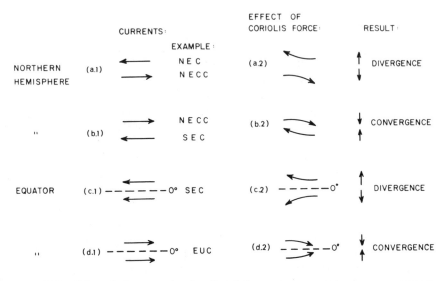

FIG. 7.28.   *Schematic diagram to show how Coriolis force acts on currents to cause convergence or divergence.*

the water. The reader may apply similar reasoning to determine what should happen between currents in the southern hemisphere (Coriolis force acting to the left of the velocity). For currents centred at the equator, flow to the west will cause *divergence* (Fig. 7.28(c) (1) and (c) (2)) as occurs at the surface in the SEC, while flow to the east will result in *convergence* (Fig. 7.28(d) (1) and d(2)). This is the main reason why the eastward-flowing EUC tends to remain close to and symmetrical about the equator during its long passage across the Pacific. The convergent tendency for the EUC below the surface gives rise, because of conservation of volume, to upward movement above the current core and downward movement below it, causing the bowing of the isopleths (Fig. 7.27(c) and (d)).

Another way to show these convergences and divergences is by means of a north–south section across the equator as displayed schematically in Fig. 7.29. In this figure, (a) shows a plan view of the trade wind pattern and the three main upper-layer currents, while (b) is a cross-section from the surface to 200 m depth to show the character of the horizontal and vertical circulations and the thermocline. Note that the slopes of the sea surface and the isotherms are grossly exaggerated (vertical/horizontal scale exaggeration about 5000 to 1).

The divergences result in upwelling of water which is often richer in nutrients than the displaced surface water and so biological production is promoted. At convergences, concentration of the upper-layer plankton may occur (because they are brought together horizontally by the flow but resist the

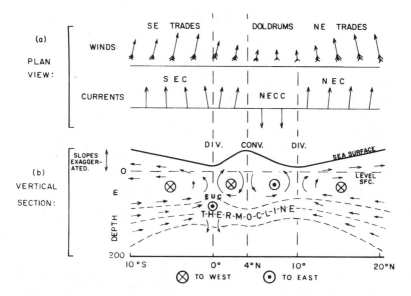

FIG. 7.29. *Schematic diagram for Equatorial Pacific: (a) plan view of winds and east–west surface currents, (b) vertical south–north section showing surface slopes (exaggerated), thermocline shape and directions of currents (arrows for currents in plane of diagram, circles with cross or dot to show westward or eastward currents respectively).*

downward motion of the water). Both the convergence and divergence zones may provide food for fish and hence be profitable fishing areas.

One major distinction between the Pacific and the Atlantic equatorial current systems is that there is no obvious major transport of upper water across the equator in the Pacific as is the case by the SEC in the Atlantic. However, Wyrtki's heat-budget analysis (Section 5.371) indicates a net export of heat from the North Pacific which must be in the form of southward advection of upper water. The location is not known but likely to be in the west and/or east rather than in mid-ocean.

7.62 North Pacific circulation

The general pattern of the North Pacific circulation is similar to that in the Atlantic with a concentrated north-eastward-flowing current on the west side while the southward return flow is spread over much of the remainder of the ocean. A difference between these two northern oceans is that the Pacific is essentially closed at the north, with just a small outflow to the Arctic and no return flow therefrom, whereas the Atlantic has a much more open connection with that sea and a considerable volume exchange with it.

7.621 North Pacific gyre and the Kuroshio

The *North Pacific Gyre* may be considered to start with the NEC flowing west and, on approaching the western boundary of the ocean, dividing with some water going south to the NEC and some north (Fig. 7.24). This water continues north-east past Japan as a concentrated current called the *Kuroshio* which is the counterpart of the Florida Current in the Atlantic. After it leaves the Japanese coast to flow east it is called the *Kuroshio Extension* to about 170° E, corresponding to the Gulf Stream and from there it is referred to as the *North Pacific Current*. The volume transport of the Kuroshio itself is about 40 Sv and of the Kuroshio Extension is about 65 Sv, comparable to those of the Florida Current and the Gulf Stream. Contributing to the North Pacific Current is the *Oyashio*, coming from the north from the Bering Sea and with some contribution from the Sea of Okhotsk. As the North Pacific Current approaches the North American continent it divides. Part turns south as the *California Current* and eventually feeds into the NEC. The remainder swings north to form the *Alaskan Gyre* in the Gulf of Alaska, and then some of it flows between the Aleutian Islands into the Bering Sea (see Dodimead *et al.*, 1963; Favorite *et al.*, 1976 for a full description). The position of the divergence of the North Pacific Current is at about 45° N in winter and 50° N in summer. The seasonal shift in this position has been clearly demonstrated by the tracks of satellite-tracked drogue buoys released in the North Pacific Current. More of the buoys which reached this divergence in summer turned north-ward than did those which arrived in winter. Many of the buoys, travelling northward into the Alaskan Gyre, became trapped in closed eddy circulations which occur off the coasts of Alaska and British Columbia. These eddies are related to features in the bottom topography (sea mounts) and may be permanent components of the circulation. Similar eddies are believed to populate the westward-flowing component of the Alaskan gyre which may account for the results of some studies showing return flow in the gyre well to the east of the Aleutians.

The apparent relation of the Kuroshio to fisheries and its probable influence on the climate of Japan and hence on the rice crop, has stimulated Japanese interest in this current, and observations of it started in the early 1900s. Modern scientific studies commenced in the early 1920s and regular monitoring in 1955 with several oceanographic cruises per year, and a number of multiple-ship studies including the international Cooperative Study of the Kuroshio (CSK) commencing in 1965 (e.g. see Stommel and Yoshida, 1972). Water properties were measured and currents observed with the GEK at the surface and with current meters and neutrally buoyant floats below the surface. Analyses of the data obtained from this and earlier studies revealed that the axis of the Kuroshio showed a distinct long-term variation in position. In Fig. 7.30 the full line represents the average position of the current during

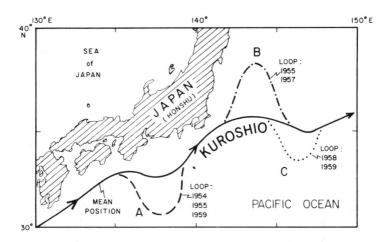

FIG. 7.30.   *Kuroshio—mean position and fluctuations during 1954 to 1959.*

1954 to 1959 but this 6-year average conceals many changes. In 1954 and 1955 a great loop to the south developed in area A (Fig. 7.30) and in the region between it and the coast an unusual cold water mass appeared. In 1956 to 1958 the current moved back to the coast but in 1959 the loop to the south developed in area A and with it the inshore cold water mass. (It persisted through 1963, straightened out in 1964 and reappeared as a loop with a cold core in 1969.) In 1955 and 1957 the current swung farther north than the average in area B, while in 1958 and 1959 it swung farther south in area C. Several reasons for these long-period changes and for the appearance of the cold-water mass have been put forward but none is considered entirely satisfactory. The maximum current speeds in the Kuroshio for the period ranged between 75 and 250 cm/s while the width of the current in which the speed was over 100 cm/s was less than 80 km most of the time. There seems to be some indication that the higher current speeds and the greater widths occurred together. If this is also characteristic of the deeper parts of the current it would imply considerable variations in the volume transport. The possible existence of a south-westward countercurrent below the Kuroshio, similar to that below the Gulf Stream, has already been mentioned.

7.63 South Pacific circulation

The *South Pacific Gyre* is less well documented than that in the North Pacific. The SEC forms the northern part of the South Pacific circulation and carries water into the Coral Sea off north-eastern Australia; in the south, the Antarctic Circumpolar Current forms the easterly flow component of the gyre. The exact character of the connection off Australia is in doubt. Although current atlases show a southward flow (*East Australian Current*) along the east coast of Australia (as in Fig. 7.24) Hamon and Golding (1980) point out that oceanographic studies of the western Tasman Sea between Australia and New Zealand have not shown a continuous current along the coast. When the south-west flows are observed they are generally accompanied by north-west flows offshore. Shipboard XBT surveys along with satellite-tracked buoy records have clearly indicated the presence of large, anticlockwise eddies of 200 to 300 km diameter with speeds of 180 to 200 cm/s and lifetimes of up to a year. The centres of the eddies are well mixed to as much as 300 m. In the austral winter, the surface water in an eddy may be as much as 2C° warmer than the surrounding water but in summer no clear temperature pattern related to the eddies may be apparent.

In the south, part of the Antarctic Circumpolar Current turns north up the South American coast as the *Peru Current*. This current has been examined by several expeditions and the coastal side of it, at least, is well described. Its volume transport is only about one-third of that of the Gulf Stream or the Kuroshio. It turns to the west near the equator and contributes to the equatorial current system. The low-temperature tongue extending west along the equator (Figs. 4.3 and 4.4) which looks like a continuation of the Peru Current was attributed by Sverdrup to upwelling of cool, subsurface water associated with the divergence to be expected with a westward current straddling the equator, as explained previously.

7.64 Eastern boundary currents; Peru Current and *El Niño*

The *Peru Current* is an example of an eastern boundary current (Wooster and Reid, 1963) which is characteristically broad and slow, contrasting with the narrow and swift western boundary currents. The Peru Current flows equatorward (Fig. 7.24) and as it comes from high latitudes it is cool and of relatively low salinity. Figure 7.31 shows vertical sections of water properties to 1000 m depth and to 1000 km from the coast at 33°S. The isopycnals have the same slopes as the isotherms shown in Fig. 7.31(a) and reveal the equatorward Peru Current above about 500 m depth and the subsurface *Gunther Current* toward the pole near the coast. The whole column is of relatively low salinity (Fig. 7.31(b)) with lowest values near the coast due to the wind-driven upwelling from a few hundred metres depth which occurs there because of prevailing equatorward winds (see Chapter 8). Another feature of the eastern boundary is the low dissolved oxygen content below the surface (Fig. 7.31(c)) near the coast brought from the tropic to the north by the coastal undercurrent. High nutrient content, represented here by inorganic phosphorus (Fig. 7.31(d)) and associated with the low oxygen, promotes the biological production which is characteristic of this eastern boundary region. These features described for the ocean off Chile and Peru are found along

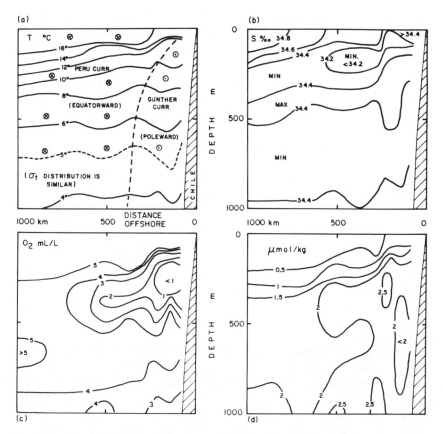

FIG. 7.31.   *Eastern South Pacific—west–east vertical profiles at 33° S to show (a) temperature with meridional current directions, (b) salinity, (c) dissolved oxygen, (d) phosphorus as phosphate.*

other eastern boundaries such as in the Benguela Current in the Atlantic off South Africa, in the Canary Current off North Africa and in the California Current in the North Pacific. Upwelling results in cooler water near the coast because temperature always decreases with increasing depth in these regions.

Generally the Peru Current extends to a few degrees south of the equator before turning west and the low temperature of its surface waters is in contrast with higher temperatures north of it. The high temperatures extend farthest south during the southern summer, December to March. At irregular intervals of a few years the high temperatures extend 5 to 10 degrees farther south than usual and the thermocline deepens by 100 m or so, the oceanographic condition known as "*El Niño*". The increase in temperature was thought to kill fish but recent studies have shown that the fish merely descend below the abnormally warm surface layer. The absence of fish in the upper layer leads to the death of many sea birds and a decrease in catch by local fishermen. The temperature increase is also accompanied by increased evaporation and consequent greater precipitation on the neighbouring land, so that floods cause much damage in this region where the normal rainfall is very small.

The early ideas about the reasons for El Niño focused on local mechanisms but during the 1970s increased interest in and observations of the phenomenon itself and of oceanic scale atmosphere/ocean interactions have suggested that more distant forces may also play a part in determining the occurrence of increases in sea surface temperature (SST) in the Peru coastal region.

The local mechanism explanation is that reduction in the equatorward coastal wind reduces the equatorward longshore wind stress, resulting in reduced upwelling of cool water; this allows the local SST to increase abnormally due to the net heat inflow through the surface in the summer. A second explanation is that the "front" between the warm, low-salinity water north of a line from the Galapagos to Ecuador and the cold, more saline Peru coastal water to the south breaks down and allows the warm Gulf of Panama water to flow south along the Peru coast. A related phenomenon is that reduction in the north-east trade winds in the eastern tropical Pacific will allow the surface waters there to warm and also result in increased eastward transport of (warm) water of the NECC into the pool off Central America. A third hypothesis, advanced by Wyrtki (1975), is that increased zonal stress to the west due to stronger than normal south-east trade winds in the open Pacific causes a piling up of warm water in the western Pacific; then, when the winds and stress relax, the water mass moves back east, strengthening the eastward NECC and EUC and causing an increase of warm upper water in the eastern Pacific and hence an occurrence of El Niño. In addition, it has been suggested that an eastward extension of the SECC might contribute warm water directly to the Peru coastal region.

A statistical study by Barnett (1977) of the relations between the atmospheric factors and the SST data suggested that (a) upwelling effects due to local (coastal) changes in wind stress were not important, (b) the eastward advection of heat by the NECC was more important to the eastern tropical Pacific than local heating and (c) about one-half of the SST variation could be predicted from sea-level changes across the Pacific, related to the third hypothesis above, suggesting again that eastward advection is very important while southward advection from the Gulf of Panama is less important (but not insignificant). This study simply tested the consistency between the hypotheses and the data but did not decide between them. In fact, it is possible that more than one mechanism may contribute to the El Niño in some degree on different occasions. One of the main difficulties at the present is that the extent of data both in time and in space is meagre, and is inadequate for making definite decisions between mechanisms. Studies of this phenomenon are continuing using both field observations and numerical modelling techniques.

It has been suggested that similar conditions to the *El Niño* probably occur annually off other upwelling coasts, such as those of California and South-west Africa, when the equatorward winds die down seasonally. However, the consequences are small compared with those along the Peru coast where a significant seasonal decrease of wind is unusual.

7.65 Pacific Ocean water masses

The water masses of the Pacific are rather more complicated than those of the Atlantic, probably because the greater size of the Pacific provides more opportunity for different masses to develop and be maintained in the different parts of the ocean (see Fig. 7.32 presented later).

7.651 Pacific Ocean upper waters

At the surface, the salinity shows the typical maxima in the tropics with a minimum in the equatorial regions (Fig. 4.9), and with lower values at higher latitudes. The surface salinity in the North Pacific is considerably less than in the North Atlantic, because of the greater runoff and precipitation. In the South Pacific the average salinity is higher than in the North Pacific but slightly less than in the South Atlantic. The surface temperature (Figs. 4.3 and 4.4) shows the usual equatorial maximum, the highest values being in the west at the down-stream end of the westward-flowing equatorial currents. There is also the previously mentioned minimum along the equator in the east attributable to the effect of upwelling. Apart from the perturbation due to the Peru Current and some effect of upwelling along the mid-latitude shores of the American continent, the distribution of the surface isotherms is distinctly zonal.

It is in the upper water masses below the surface layer that the variety of water properties is found. Between depths of about 100 and 800 m there are the *North* and *South Pacific Central* and the *Pacific Equatorial Water Masses* (Fig. 7.13). The Equatorial Water extends almost from shore to shore, but despite its extent of some 12,000 km it has a very uniform T–S relationship across the width of the ocean. This is evident in the mean T–S curves for 5° squares for the mid-Pacific (Emery and Dewar, 1981) shown in Fig. 7.32. These are similar to those for the Atlantic (Fig. 7.16) with a central mean T–S curve and one standard deviation curves on either side of the mean. (The low salinities in the Gulf of Panama referred to earlier are very apparent in this figure.) In the eastern Pacific the Equatorial Water extends from about 20° N to 10° S but it diminishes in north–south extent to the west. Except for the waters in the western South Pacific it is the most saline water in this ocean, but even so is slightly less saline than the Atlantic Central waters (Fig. 7.14). The Equatorial Water is separated from the well-mixed and homogeneous surface

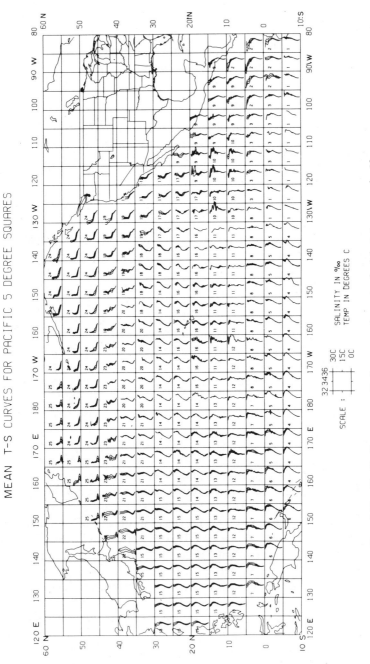

FIG. 7.32. *Pacific Ocean: mean T–S curves and one standard deviation curves by 5° squares.*

layer by a very strong thermocline, and vertical transfer of water properties up or down is inhibited by the stability of this water. For this reason it is often referred to as the "discontinuity layer". The depth of the discontinuity layer decreases from 150 to 200 m in the west to 50 m or less in the east; the layer even seems to reach the surface at times near the American coast. This effect is associated with the *Costa Rica Thermal Dome*, an apt name which describes the three-dimensional shape of the thermocline with the cold water of the lower thermocline occasionally penetrating to the sea surface itself.

At about the level of the thermocline in the Equatorial Waters is the salinity maximum which is characteristic of the equatorial regions (Fig. 4.11). It is of limited areal extent and so does not appear in Fig. 7.13 which shows averages over large areas and it is of too small vertical extent to be visible in the T–S curves of Fig. 7.32. At about 800 m there is a slight salinity minimum which represents the limit of the northward influence of the Antarctic Intermediate Water (Fig. 7.34).

In the Pacific between 30° N and 20° S there are two cores of remarkably low dissolved oxygen values between 200 and 1000 m (Fig. 4.12). These cores are centered at 10° N and 7° S and extend westward right across the Pacific, the values increasing from less than 0.1 mL/L in the east to 3 mL/L in the west. Figure 7.33 shows these oxygen minima plotted on the surface where $\Delta_{S,T}$ = 160 × 10^{-8} m^3/kg (σ_t = 26.5). At the centre of the oxygen minimum cores, this surface is at a depth of about 150 m at the extreme east, descending to 250 m in the west. Below these oxygen minimum layers in the upper waters the values rise again in the deep water. Corresponding to the general low oxygen values are high concentrations of nutrients, particularly in the North Pacific. Values of 2 to 3.5 μmol/kg phosphorus as phosphate are typical in the North Pacific compared with values of 0.5 to 2 μmol/kg in the Atlantic. Similar large concentrations of dissolved nitrate and silicate are also found in the North

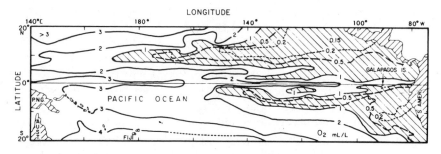

FIG. 7.33. *Equatorial Pacific Ocean—dissolved oxygen values on surface* $\Delta_{S,T} = 160$ × 10^{-8} m^3/kg. *(Approximate depths of this surface are: at 1 mL/L about 200 m; 2 mL/L about 250 m; 3 mL/L about 300.)*

Pacific waters. These materials go into solution during the decomposition of organic matter, e.g. dead plankton. High concentrations may be a result of a copious supply of organic remains from life in the upper waters, or may accumulate as a consequence of the slow movement of the deep water. In the case of the Pacific, the latter is assumed to be the main cause.

The North Pacific Central Water extends from the Equatorial Water to about 40° N and is the least saline of the central water masses of the oceans. Sverdrup distinguished an extensive western North Pacific Central Water to the west of the Hawaiian Islands from a smaller eastern North Pacific Central Water between these islands and the North American coast, though separated from the coast by another water mass. Examination of the accumulation of data for the North Pacific since 1955 has indicated that the two water masses are less easily distinguished now than they were with the lesser amount of data available to Sverdrup. They are not separated in Fig. 7.13(a).

North of the North Pacific Central Water is a *Pacific Subarctic Water Mass* (Favorite *et al.*, 1976) which extends across the greater part of the ocean. Its characteristic properties are low salinity (33.5 to 34.5‰) and relatively low temperature (2 to 4°C). It is a much more extensive and important water mass than the Atlantic Subarctic water. It corresponds to the Subantarctic Water in the south but its mode of formation must be somewhat different. The Subantarctic Water is formed between the Subtropical Convergence and the Antarctic Polar Front which are essentially continuous across the South Pacific. In the North Pacific the *Subarctic Convergence* is only clear in the western part of the North Pacific (Fig. 7.24) where the Subarctic Water must be formed by mixing between the warm, saline waters of the Kuroshio Extension and the cold, less saline waters of the Oyashio. In the eastern North Pacific there is certainly a gradation from Subarctic to Central Water but it is less easy to localize than is the Antarctic Polar Front. The Subarctic Water forms a part of the North Pacific Current flowing to the east and it retains its characteristics of low salinity and temperature until it approaches the American coast. Here part of it swings south-east and its temperature starts to rise by heating and the salinity by mixing until it attains the typical Equatorial Water characteristics as it merges into the North Equatorial Current.

South of the Equatorial Water are the western and eastern *South Pacific Central Water Masses* extending to the Subtropical Convergence at 40°S which is usually considered as the oceanographic southern boundary of the Pacific Ocean. In the South Pacific the Central Water Masses are less easily distinguished from the Equatorial Water than is the case in the North Pacific.

Intermediate water masses are found below the Central Waters in both the North and South Pacific (Reid, 1965, 1973). In the latter, the *Antarctic Intermediate Water* is formed by subsurface mixing at and north of the Antarctic Polar Front with fairly well-defined properties, a temperature close to 2.2° C and a salinity of 33.8‰. This salinity is relatively low and so the

Antarctic Intermediate Water gives rise to the minimum in the T–S diagram below the Subantarctic and Central Waters (Fig. 7.13(b)). The Intermediate Water flows north as seen in Fig. 7.34, increasing in salinity by mixing with waters above and below but is limited in northward extent by the Pacific Equatorial Water. This contrasts with the situation in the Atlantic where, in the absence of a clearly defined Equatorial Water Mass, the Antarctic

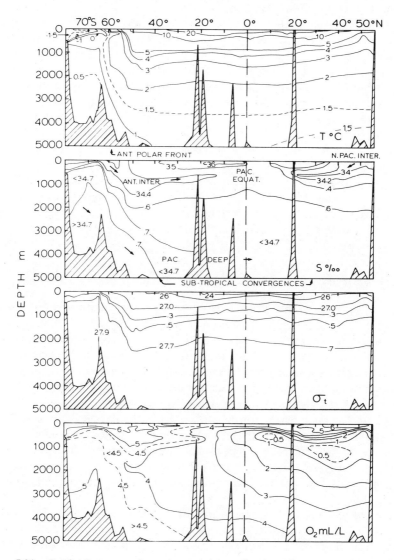

FIG. 7.34.   *Pacific Ocean—south–north vertical sections at 160° W of water properties: (a) temperature, (b) salinity, (c) sigma-t, (d) dissolved oxygen.*

Intermediate Water continues north across the equator. In the North Pacific a *North Pacific Intermediate Water*, evident as a salinity minimum, is found below the North Pacific Central Water (Figs. 7.13 and 7.34). In the west it is deepest at about 800 m but rises to about 300 m in the east. It appears to circulate in a clockwise gyre, similar to the surface water, and its relatively high oxygen content indicates frequent replenishment. Sverdrup suggested that it is formed in the west near Japan by sinking at the convergence of the Oyashio and the Kuroshio Extension. Reid (1965) pointed out that the salinity minimum of the North Pacific Intermediate Water occurs on an isopycnal surface of $\sigma_t = 26.8$ and that water of this density very rarely occurs at the surface in the North Pacific. Where this density is found at the surface in the South Pacific (as a possible alternative source) the salinity is higher than in the salinity minimum of the North Pacific Intermediate Water. Reid therefore concluded that this latter water attains its properties below the surface by vertical mixing at the north-west in the Subarctic region where the particular density surface is shallow and the surface waters above are cold, low in salinity and high in oxygen content. It was thought that this was one of the few exceptions to the general rule that subsurface water masses acquire their T–S characteristics at the surface of the sea, but evidence is growing that many recognized water masses are formed below the surface by the mixing of a lower layer with an upper layer of different properties which is itself in contact with the atmosphere.

7.652 Pacific Ocean deep waters

The deep water of the Pacific is characterized by very uniform properties as may be seen in the meridional sections of Fig. 7.34. Between 2000 m and the bottom the temperature range (as far south as the Subtropical Convergence) is only from 1.1 to 2.2° C and the salinity from 34.65 to 34.75‰. The salinity tends to increase with depth or to remain constant, in contrast to the situation in the Atlantic where there is a marked salinity maximum of the North Atlantic Deep Water at mid-depth with a decrease to lower values in the Bottom Water (Fig. 7.9). The main reason for the uniformity in the Pacific Ocean is that no deep water is formed in it. In the north, the salinity is too low for winter cooling to make the water dense enough to sink to any considerable depth. In the south, although some bottom water probably comes from the Ross Sea, the volume is not large and it appears to be absorbed first into the circumpolar circulation.

The *Pacific Deep* and *Bottom Water*, from 2500 m down, must come from the other oceans, entering from the south-west between New Zealand and Antarctica. It spreads north at 180° to 160°W and, after crossing the equator, branches north-west and north-east along basins into the North Pacific. The *Scorpio* Expedition of 1967 (e.g. Warren, 1973), consisting of two lines of stations right across the Pacific at 28° and 43°S, was designed to investigate the deep waters more thoroughly than before. One particular result was that the distribution of properties indicated a narrow flow between 3000 and 4000 m depth at 28°S along the east side of the Tonga–Kermadec Ridge (about 175°W), northward into the Pacific. The transport was calculated as about 10 Sv compared to some 14 Sv required for a steady-state budget for heat, salt, volume and ^{14}C. The evidence for the clockwise flow of the bottom water round the Pacific is the gradual increase of temperature along the route, from about 0.9°C in the Antarctic to 1.5 to 1.6°C in the eastern North Pacific (Mantyla, 1975), and a small decrease in oxygen content (Fig. 7.34). There is a slight indication of a decrease in salinity into the North Pacific but the change is close to the limit of measurement precision. The rise of temperature is attributed to heat flow from the interior of the earth into the sea. The water which penetrates into the North Pacific must rise into the upper layers above 2000 m.

Sverdrup considered, essentially by analogy with the Atlantic, that a slow southward movement of deep water (i.e. between the Bottom Water and the Intermediate Water) must occur in the South Pacific. This southward movement of deep water would be supplied from the northward movement of Bottom Water below it and the Intermediate Water above. The indications are, therefore, that the deep and bottom water movements in the Pacific are very slow and that there is only a small exchange of water between the North and South Pacific in contrast with the large volume exchange at various levels across the equator in the Atlantic.

The water masses of the Pacific and Indian Oceans are very similar and in his analysis of the distribution of temperature and salinity Montgomery (1958) pointed out that this water, with its mean temperature of 1.5°C and salinity of 34.70°/$_{oo}$, forms the largest water mass in the world ocean. For this reason he named it the (Oceanic) Common Water from which other water masses are distinguished by differences of temperature and salinity. The Common Water has average properties because it is a mixture of other water masses, chiefly North Atlantic Deep and Antarctic Bottom Waters with some admixture from the Indian and Pacific Oceans.

Although the large body of deep water in the Pacific is usually called the Pacific Deep Water, the name in this case indicates the region where the mass is found, rather than where it is formed as is usually the case with water mass names.

Comparison of Fig. 7.34 with Fig. 7.9 will show that in addition to differing from the Atlantic in being more uniform in temperature and salinity distributions, the Pacific water also has lower dissolved oxygen values. In the Atlantic the values range from 3 to 6.5 mL/L in the deep water, whereas they are from 0.5 to 4.5 mL/L in the Pacific. These lower values are consistent with the presumed greater age and slower circulation of the deep water in the

Pacific, or perhaps it would be better to say that they are one of the reasons for believing the circulation to be slow.

One feature which is lacking in the Pacific, as compared with the Atlantic and the Indian Oceans, is a source of high salinity, warm water such as comes from the European Mediterranean and from the Red Sea respectively. These water masses are not of major importance in terms of volume but are very useful as tracers of the movement of subsurface waters.

7.7 Indian Ocean

The Indian Ocean differs from the Atlantic and Pacific Oceans in its limited northward extent, to only 25°N. The southern boundary of the ocean is usually taken at the Subtropical Convergence at about 40°S.

The Indian Ocean used to be one of the least known oceans but the extensive observations made during the International Indian Ocean Expedition from 1962 to 1965 increased our knowledge immensely, even though investigations emphasized the interesting equatorial zone at the expense of the inhospitable southern part of the Ocean. After this Expedition, Wyrtki brought together the data from that cooperation, and also all available earlier material, in the *Oceanographic Atlas of the International Indian Ocean Expedition* (Wyrtki, 1971) in which the data are displayed in various ways to show seasonal as well as average distributions, and relations between properties in a variety of characteristic diagrams. The following description is largely based on Wyrtki's Atlas and his article (1973).

7.71 Indian Ocean circulation

It is in the wind-driven equatorial current system and in its northern parts that the Indian Ocean circulation differs most from those in the Atlantic and Pacific. Because of the land mass to the north of the ocean there is a seasonal variation in the winds north of the equator (Fig. 7.25). From November to March, these winds blow from the north-east (north-east Trades or *North-east Monsoon*), from May to September they blow from the south-west (*South-west Monsoon*). (The word *monsoon* is derived from an Arabic word meaning winds which change with the seasons.) The south-west monsoon winds are really a continuation across the equator of the south-east Trades which continue throughout the year. The change of wind direction north of the equator then results in a change of currents there. During the north-east monsoon (November to March) there is a westward-flowing *North Equatorial Current* from 8°N to the equator; from the equator to 8°S there is the eastward-flowing *Equatorial Countercurrent* and from 8°S to between 15° and 20°S there is the *South Equatorial Current* to the west (Fig. 7.35(a)). During the south-west monsoon (May to September) the flow north of the equator is reversed and is to the east. This combines with the eastward ECC and the

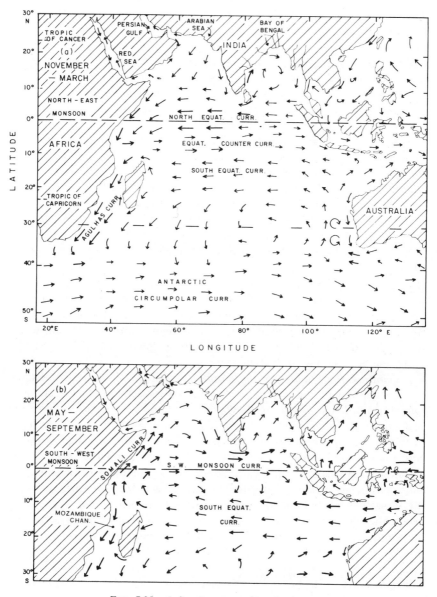

FIG. 7.35. *Indian Ocean—surface circulations.*

whole eastward flow from 15°N to 7°S is called the *(South-west) Monsoon Current* (Fig. 7.35(b)). The SEC continues to the west south of 7°S but is stronger than during the north-east monsoon.

An *Equatorial Undercurrent* is found in the thermocline east of 60°E during the north-east monsoon period. It is weaker than those in the Pacific and

Atlantic. During the south-west monsoon, with the general flow to the east at the equator, the EUC is not evident. Studies of surface currents from ship observations and from oceanographic data indicate that the changeover from one monsoon regime to the other takes place in 4 to 6 weeks. This is one of the few places in the oceans where such marked changes in wind take place and the information on the rate of response of the current pattern has been very valuable to dynamic oceanographers requiring information on rates of change of currents in response to changing forces.

The currents at the African shore are interesting. During the north-east monsoon period (November to March) the SEC, when it reaches the African shore, supplies both the ECC to its north and the *Agulhas Current* flowing south (Fig. 7. 35(a)). This current is deep and remarkably narrow, probably only 100 km wide, and flows south close to the African shore with a transport reported to average 50 Sv and to rise to 80 Sv at times. When it reaches the southern tip of Africa, the current turns east into the Circumpolar Current. During the south-west monsoon (May to September), the component of the SEC which turns north supplies the *Somali Current* up the east coast of Africa (Fig. 7. 35(b)). This current is notable for its high speeds of up to 200 cm/s, which are comparable to those of the Florida Current, and it has a transport of about 65 Sv, most of it in the upper 200 m. The South Equatorial Current, the Somali Current and the Monsoon Current then comprise a strong wind-driven gyre in the northern Indian Ocean. Strong upwelling occurs at this time along the Somali and Arabian coasts.

Along the eastern boundary of the Indian Ocean, off Western Australia, the flow contrasts with that in other oceans in two respects. Satellite tracking of buoys (Hamon and Goulding, 1980) shows the flow to be generally to the south (although atlases show northward flow) and also there is little evidence of upwelling along this eastern boundary. In addition, there are mesoscale eddies which are both cyclonic and anticyclonic, in contrast to those off the eastern Australian coast which are anticyclonic only. In the austral autumn, the *Leeuwin Current* of warm, low salinity water flows southward down the coast off Western Australia and then east into the Great Australian Bight. This occurs during the transition from the south-west monsoon to the north-east monsoon, but the reason for this short duration current is not known.

7.72 Indian Ocean water masses

The surface water masses in the open Indian Ocean have typical characteristics, a substantially zonal distribution of the isotherms with a temperature maximum near the equator (Figs. 4.3, 4.4) and a salinity maximum at about 30°S in the eastern ocean (Fig. 4.9). This is somewhat farther south than in the Atlantic and Pacific. In the north, the *Arabian Sea* west of India and the *Bay of Bengal* to the east have very different characteristics. The Arabian Sea has high surface salinity values up to 36.5°/$_{oo}$, due to evaporation, while in the Bay of Bengal the salinity decreases from about 34°/$_{oo}$ at about 5°N to 31°/$_{oo}$ or less in the north. The low values in the Bay of Bengal are due to the very considerable river runoff into it, particularly during the south-west monsoon. Some of this low-salinity water is carried westward while heavy rainfall in the Intertropical Convergence Zone contributes to the low salinity in the northern part of the SEC.

Below the surface layer (Fig. 7.14(b)) and north of 10°S there is the *Indian Equatorial Water* or *North Indian High-salinity Intermediate Water* (Wyrtki, 1973). This water has a relatively uniform salinity of 34.9 to 35.5°/$_{oo}$. Some of it is formed in the Arabian Sea but there are also components from the Red Sea and the Persian Gulf with values to 36.2°/$_{oo}$. It also spreads into the Bay of Bengal. Wyrtki points out that the waters of the northern monsoon gyre are separated from the (southern) subtropical waters by a boundary or front which is inclined from about 100 m depth at 10°S to 1000 m depth at 20°S. North of this front, the water is saline, low in oxygen and high in nutrients while to the south the reverse is the case. South of the equator to about 40°S and to 1000 m is the *Indian Central Water* of the southern gyre of the ocean with salinities from 34.5 to 36°/$_{oo}$; under this is the Antarctic Intermediate Water. Both the *Deep* and *Bottom Waters* are of Atlantic/Antarctic origin, as no waters of these types are formed in the Indian Ocean. As in the Pacific Ocean the circulation of these waters appears to be slow and must be inferred from the property distributions. The Deep Water properties are 2°C and 34.8°/$_{oo}$, essentially those of Montgomery's "Common Water". A flow of Bottom Water to the north in the west and east basins (separated by the mid-ocean ridge) is inferred from a slight increase in potential temperature to the north attributed to geothermal heating from below. A very noticeable feature of the deep waters is the decrease in dissolved oxygen from 5 mL/L in the Antarctic waters to 0.2 mL/L in the Arabian Sea and Bay of Bengal. This is attributed to the recent formation and continued movement of the southern waters contrasted with the isolation and stagnation in the north. This situation is in contrast with that in the Atlantic and Pacific where the most notable regions of low oxygen are on the east side around the equator.

7.8 Red Sea and Persian Gulf

The depths in the *Red Sea* average about 1000 m with basins to over 2200 m, while there is a sill of about 110 m depth toward the south end. Because of the very dry climate, evaporation exceeds precipitation by 200 cm/year and there is no river runoff into the Red Sea. In consequence, high salinities occur, up to 42.5°/$_{oo}$ at the north end, combined with high temperatures up to 30°C in summer and 18°C in winter in the surface waters. The main body of *Deep Water* below sill depth is very uniform at 21.7°C and

$40.6°/_{oo}$. This water is formed in the winter at the north end of the Sea. Neumann and Gill (1961) presented evidence to show that the water which flows out of the Red Sea into the Indian Ocean is not the deep basin water, but water recently cooled at the surface and flowing to the mouth in a shallow layer at and above the 110-m sill depth. As a result, the *Intermediate Water* which leaves the Red Sea is rather different in properties from the Red Sea Deep Water. Inside the sill the Intermediate Water is at about 24°C and $39.8°/_{oo}$, but during its flow over the sill into the Indian Ocean mixing reduces these values to 15°C and $36°/_{oo}$. The subsurface outflow is, of course, compensated by a surface inflow from the Indian Ocean. This exchange with the outside ocean is very similar to that which Wüst described for the Mediterranean, particularly in respect of the outflow being from the shallow Intermediate Water rather than due to displacement of the deep basin water. The Red Sea outflow is a readily identified feature of the Arabian Sea and northern Indian Ocean. It penetrates south along the west side of the ocean at about 600 m to 25°S in the Mozambique Channel.

Notable features of the Red Sea are the hot brine pools found in some of the deepest parts (e.g. see Degens and Ross, Eds., 1969). Temperatures of 58°C and salinities of $320°/_{oo}$ have been recorded, although this latter figure is not directly comparable to ocean water salinities as the chemical constitution of these brines is very different. They have a much higher content of metal ions. (For comparison, a saturated solution of sodium chloride in water has a salinity in the oceanographic sense of about $270°/_{oo}$.) Various explanations have been offered for the origin of these brine pools. The one with the fewest arguments against it is that this is interstitial water from sediments, or solutions in water of crystallization from solid materials in the sea bottom, released by heating from below and forced out through cracks into the deep basins of the Red Sea. In this connection it should be noted that the Red Sea is one of the two places where a mid-ocean ridge runs into a continent, and it is a region where the heat flow up through the bottom is much greater than the world average of $4 \times 10^{-2} W/m^2$.

In contrast to the Red Sea, the *Persian Gulf* has a mean depth of only 25 m. Water of temperature between 15° and 35°C and up to $42°/_{oo}$ is formed here and some is contributed to the Arabian Sea.

CHAPTER EIGHT

Coastal Oceanography

8.1 Introduction

Oceanographic conditions in coastal waters differ in many respects from those in the open sea. Some of the factors causing these differences are the presence of the coast as a boundary to flow, the shallowness of the water over the continental shelf, river runoff and precipitation, and the effects of continental air masses flowing out over the sea. In particular, variations of water properties and motions with position and with time are larger than is generally the case in the open ocean. Many of the world's fisheries are in coastal waters, and other problems such as those concerned with the disposal of sewage or industrial effluent are of immediate importance. Some of the characteristics of coastal oceanography will therefore be described.

The first effect of the shore as a *boundary* is obvious – it limits possible directions of motion so that horizontal flows tend to parallel the coast. It is important to recognize this effect because it represents one of the few situations in which man can exert a significant influence on the ocean. Jetties and break-waters, designed for the protection of shipping from swell, may also redirect currents. In the past this has often led to unexpected changes such as the deposition of silt and the formation of new shoals, or the removal of beaches when a jetty prevented the longshore movement of sand required to maintain a beach whose material was being eroded away by wave action.

Less immediately obvious are the effects of the coast in causing vertical motions of the water. One of these is the vertical motion which we call the *tide*, i.e. the semidiurnal or diurnal rise and fall of sea level due to the coast blocking the horizontal flows generated by the periodically varying astronomical forces and causing the water to pile up against the coast during the flood or fall away during the ebb. The other effect results when surface water is forced away from a significant length of coast; the presence of the latter prevents replacement by other surface water and, instead replacement takes place from below, called *coastal upwelling* (as distinct from the upwelling which takes place between currents in the open ocean as described in Chapter 7). This process will be discussed in the next section.

In shallow areas, e.g. over the continental shelf, there is a limited reservoir of deep water for mixing and, in consequence, the input of solar energy can cause a greater summer rise of temperature than in the open ocean.

The effects of *tidal currents* are twofold. They may cause large changes twice daily in the volume of water in a harbour or bay, and they may also promote vertical mixing and thus break down the stratification of the water. For instance, where there are strong tidal currents over a rough bottom in shallow water the turbulence causes the heat absorbed in the surface layer to be mixed through a considerable depth. A result will be lower surface temperature changes compared with those toward the head of a bay nearby where the currents are slight.

The effect of cold, dry, continental *arctic air* flowing out over the ocean and enhancing the heat loss in the north-east Atlantic and Pacific has been discussed in Chapter 5. In other areas, dry air from desert regions may promote evaporation and may also carry terrestrial material as dust to the open ocean.

The direct effect of *river runoff* is to reduce the salinity of the surface layers, and even of the deeper water if there is sufficient vertical mixing. Generally, river runoff has a pronounced seasonal variation and this gives rise to much larger seasonal fluctuations of salinity in coastal waters than in the open ocean. In a coastal region where precipitation occurs chiefly as rain, the seasonal salinity variation will follow closely the local precipitation pattern. In regions where rivers are fed by melt water from snowfields or glaciers, the river runoff increases in the summer to many times the winter rate and causes a corresponding decrease of salinity during this period which lags the snowfall by several months. Since river water frequently carries suspended sediment this causes the coastal waters to have low optical transparency. Sometimes this sediment is carried in the surface low-salinity layer for some distance while the deeper, more saline, water remains clear. The deposition of this sediment causes shoaling and consequent hazards to navigation. Frequently the location of this deposition is influenced by the salinity distribution because increases in salinity can cause flocculation of the sediment and rapid settling. The effect of runoff can often be traced a long way from the coast, both by reduced salinity and by the sediment in the water. Examples of this are found in the Atlantic off the Amazon River, in the north-east Pacific from the many rivers flowing in from the North American continent and in the Bay of Bengal from the large rivers whose runoff is strongly influenced by the monsoon.

Another consequence of reduced salinity in coastal waters is increased stability in the halocline so that summer input of solar heat is retained in the surface layer with resulting high temperatures. In high northern latitudes, ice tends to form first in coastal waters during the cooling season because of the limited reservoir of heat in the shallow waters combined with the low salinity from river runoff.

In low latitude subtropical regions, where precipitation is small, evaporation becomes important and high salinities as well as high temperatures occur in bays and partially surrounded bodies of water.

Along the continental shelf, variations in water conditions may be important to fish populations. Sometimes these variations are seasonal in character, but important consequences of longer-term variations have also been recorded. For instance, cod appear to have a limited tolerance to water temperature and the yield of the cod fishery off West Greenland has been shown to be very sensitive to this factor. A great improvement in the yield occurred from 1924 on, when the temperature rose from below 0°C to 1°C. Previously, rich fisheries had been experienced from 1845 to 1850 when there was less ice than usual and presumably higher temperatures (water temperatures are not available for that time), but the yield had declined thereafter.

A type of coastal zone which has received little attention from physical oceanographers in the past has been the *lagoon* within the reef of coral atolls or between the coast and a barrier reef. In recent years, work has been started by Australian oceanographers in the Great Barrier Reef lagoon and by French oceanographers in the lagoon around New Caledonia.

When carrying out oceanographic surveys in coastal waters it is necessary to take into account the special characteristics of these regions. Because of the large variations of water properties over small distances it is necessary to arrange oceanographic stations closer together than in the open ocean. For instance, in the open ocean a minimum spacing of 50 to 100 km may be adequate for general surveys but near the coast it may be necessary to reduce this to 5 or 10 km or even less. (Of course, when studying small eddies or features like fronts or the edge of the Gulf Stream, close spacing will be needed even in the open ocean.) In addition, seasonal variations are much greater near shore and in order to obtain a complete picture of oceanographic conditions it is necessary to extend observations over one or more full years. In doing this a number of oceanographic cruises are necessary because single winter and summer cruises will usually be insufficient to describe the full range of water characteristics and the times and extent of changeover periods between winter and summer extremes. Because tidal currents may play an important part, it is desirable to plan time-series observations over one or more tidal days (25 hours) at key locations. It is not uncommon to find as large variations in 25 hours at a single location in coastal waters as occur over distances of 50 to 100 km simultaneously in the same region. When commencing a study of a previously undescribed region it is advisable to plan a number of such time-series studies ("anchor stations") early in the project.

8.2 Coastal Upwelling

References have been made earlier to upwelling taking place along the eastern boundaries of the oceans. A qualitative explanation of this process will

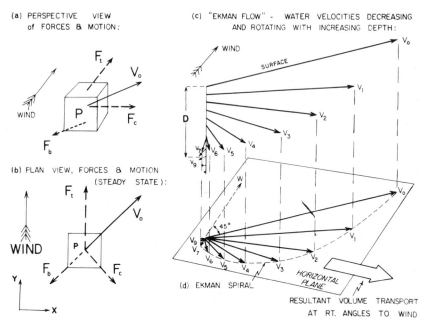

FIG. 8.1.   *Wind-driven currents—northern hemisphere: (a) forces and motion of surface parcel P, (b) plan view of forces and motion, (c) water velocity as a function of depth, (d) Ekman spiral.*

be presented. It was shown by Ekman originally (1905) to result from the effect of the rotation of the earth on bodies moving relative to the earth. Imagine a parcel of water P (Fig. 8.1 (a)) in the surface layer. When the wind W blows over the surface, it causes a frictional stress F_t in the direction of the wind on the top of the parcel which starts to move in the wind direction. Immediately the Coriolis force F_c acts at right angles to the direction of motion, tending to cause the flow V_o to be redirected to the right of the wind direction (in the northern hemisphere, to the left in the southern). At the same time, as the surface parcel commences to move a retarding friction stress develops between it and the next layer below. This retarding force F_b on the bottom of the parcel is opposite to that of the water motion V_o. In the simple application of the theory, a steady state develops in which the surface water velocity V_o is at 45° to the right of the wind direction W in the northern hemisphere (45° to the left in the southern hemisphere) (Fig. 8.1(b)). A similar argument can be applied to successive layers beneath the surface layers and each of these will move at a slower speed and in a direction further to the right of V_o. In a simulated three-dimensional view (Fig. 8.1(c)) the arrows represent the velocities in successively deeper layers, showing how the speed (length of arrow) decreases with increased depth and how the direction changes clockwise (in the northern hemisphere). If the arrows are projected on to a horizontal plane their tips form the *Ekman Spiral* (Fig. 8.1(d)). The depth D_E at which the water velocity is opposite to the surface velocity V_o is called the *depth of frictional influence* and is taken as a measure of the depth to which the surface wind stress affects water motion. This upper layer is referred to as the *Ekman Layer*. The value of D_E depends on the eddy friction in the water and on latitude (through the Coriolis parameter). Making some simplifying assumptions about the character of the eddy friction, for a wind speed of 10 m/s at latitudes $10°$, $45°$ and $80°$, the values of D_E would be 100, 50 and 45 m respectively. For a wind speed of 20 m/s, the values of D_E would be 200, 100 and 90 m respectively. (For a more detailed discussion see Pond and Pickard, *Introductory Dynamic Oceanography*.)

It can be shown that the resultant volume transport in the Ekman layer is at right angles to and to the right of the wind direction W in the northern hemisphere as shown by the large arrow in Fig. 8.1(d) (to the left in the southern hemisphere). It is this resultant flow which gives rise to upwelling. If the wind blows parallel to the coast and toward the equator at the eastern boundary of an ocean, water in the Ekman layer will tend to move *away from the coast* in either hemisphere and will have to be replaced by water upwelling from below the layer. Since the water temperature decreases as depth increases in these regions, the upwelled water is cooler than the original surface water and a characteristic band of low temperatures develops close to the coast. This may be seen off the Pacific coast of North America in Fig. 4.4 and off the Atlantic coast of South Africa in Fig. 4.3. Often the upwelled water also has greater concentrations of nutrients (phosphates, nitrates, silicates, etc.) than the original surface water which had been depleted by biological demands. The result is that upwelling is important in replenishing the surface layers with these components which are needed for biological production. This process is particularly important for the fishery off the South American coast.

It is important to note that the water which upwells is not deep water. Comparison of the properties of the upwelled water with those in the water column before the start of upwelling has shown that the water comes to the surface from depths which are usually between 50 and 300 m. It was for this reason that the earlier reference was to the upwelling of "subsurface" water only. Off the North American coast from British Columbia to California (i.e. $55°$ to $30°N$) relatively cool water is usually found from April to August within a region 80 to 300 km wide from the shore (Fig. 4.4). The cold water is often patchy, rather than being in a uniform band, and comes from depths of no more than 300 m. Off the South American coast, upwelling has been observed between $5°$ and $35°S$ and comes from an average depth of 130 m with a maximum of 350 m.

In addition to cooling the upper layer, the upwelling normally causes a decrease in salinity because in most areas salinity decreases with increased depth. The exception is off the North American coast where salinity increases with depth (the north-east Pacific being a region of low surface salinity because of river runoff and precipitation) and so upwelling increases surface salinity.

The earlier studies of upwelling areas were made off South-west Africa and in the Pacific off South America. Because of the importance of the upwelling process to the fisheries, a new study was started in 1971. The overall goal is to understand the Coastal Upwelling Ecosystem (CUE, e.g. see Barber, 1977) well enough to be able to predict the future response of the system by monitoring a few key ocean parameters. The physical studies are concerned with the quantitative response of the eastern boundary waters to the wind stress. The main areas for intensive study have been off the west coasts of North and South America, particularly off Oregon and Peru respectively, in the eastern North Atlantic and a major operation was carried out off the north-west coast of Africa in 1974. Earlier studies were reviewed by Smith (1968) and a more recent study by Bryden (1978).

8.3 Estuaries

Oceanographically the term *estuary* has a wider meaning than the conventional one of the tidal region at the mouth of a large river. Cameron and Pritchard (1963) define an estuary as "a semi-enclosed coastal body of water having a free connection to the open sea and within which the sea-water is measurably diluted with fresh water deriving from land drainage". They restrict the definition to coastal features and exclude large bodies of water such as the Baltic Sea. The river water which enters the estuary mixes to some extent with the salt water therein and eventually flows out to the open sea in the upper layer. A corresponding inflow of sea-water takes place below the upper layer. The inflow and outflow are dynamically associated so that while an increase in river flow tends to reduce the salinity of the estuary water it also causes an increased inflow of sea-water which tends to increase it. Thus an approximate steady state prevails.

8.31 Types of estuaries

In terms of shape, Pritchard (1952, also Dyer, 1973) distinguishes three types, the coastal plain, the deep basin and the bar-built estuary. The first of these is the result of land subsidence or a rise of sea-level flooding a river valley; examples are the St. Lawrence River valley and Chesapeake Bay in North America. Typical examples of the second type are the fjords of Norway, Greenland, Canada, South America and New Zealand. Most of these have a sill or region toward the seaward end which is shallower than both the main basin of the fjord and the sea outside and so restricts the exchange of deep water. The third type is the narrow channel between the shore and a bar which has built up close to it by sedimentation or wave action. The inland end of an estuary is called the *head* and the seaward end the *mouth*. Positive estuaries have a river or rivers emptying into them, usually at the head, and it is this which gives rise to the characteristic features of water property distributions in such estuaries.

Estuaries have been classified by Stommel (see Dyer, 1973) in terms of the distribution of water properties as (a) vertically mixed, (b) slightly stratified, (c) highly stratified and (d) salt wedge estuaries. The stratification referred to is of salinity, because it is typical of estuaries that the density of the water is determined mainly by salinity rather than by temperature. The salinity distributions in these four types are shown schematically in Fig. 8.2 in two ways. In the left-hand column of graphs the salinity distributions are shown as vertical profiles at each of four stations between the head and the mouth of the estuary as shown in the schematic plan view at the top of the figure. The right-hand column shows simplified longitudinal sections of salinity from head to mouth for the full depth of the estuary.

The *vertically mixed estuary* (type A, Fig. 8.2) is generally shallow and the water is mixed vertically so that it is homogeneous from surface to bottom at any particular place along the estuary. The salinity increases with distance along the estuary from head to mouth. The river water in such an estuary flows toward the mouth while the salt may be considered to progress from the sea toward the head by eddy diffusion at all depths. In the right-hand figure, the vertical isohalines indicate the homogeneity of the water at each location while the straight arrows indicate that the direction of net flow of the water is seaward at all depths. (The circular arrows symbolize the mixing taking place at all depths.) The Severn River in England is an example of such a vertically mixed estuary.

In the *slightly stratified estuary* (type B), which is usually also shallow, the salinity increases from head to mouth at all depths. The water is essentially in two layers with the upper layer a little less saline than the deeper one at each position along the estuary, with a mixing layer between them (symbolized by the circular arrows in Fig. 8.2B). In this type of estuary there is a net seaward (outward) flow in the upper layer and a net inward flow in the deeper layer as shown by the straight arrows in the vertical salinity section. In addition to these flows at both levels there is the vertical mixing of both fresh and salt water giving rise to the longitudinal variation of salinity in both layers. The James River in Chesapeake Bay is an example of this type.

In the *highly stratified estuary* (type C), of which the fjords are typical, the upper layer increases in salinity from near zero in the river at the head to a value close to that of the outside sea at the mouth. The deep water, however, is of almost uniform salinity from head to mouth. Again there is a net outflow in the upper layer and inflow in the deeper water as shown by the straight arrows

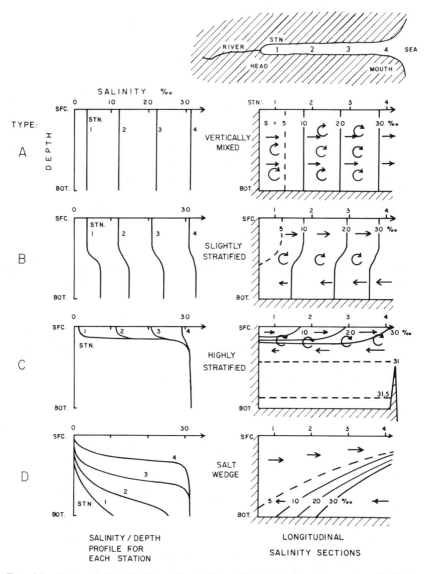

FIG. 8.2. *Typical salinity/depth profiles (left) and longitudinal salinity sections (right) in estuaries (schematic).*

in the salinity section. In these estuaries there is a very strong halocline between the upper water and the deep water, particularly at the head where vertical salinity gradients of 10 to 20‰ per metre may occur in summer during the period of greatest river runoff. There is vertical mixing but this results predominantly in an upward movement of salt water from below into the upper layer, with little downward movement of fresh water. One explanation for this almost unidirectional mixing is that internal waves are generated by the velocity shear between the upper low salinity layer and the deeper more saline water, and that the tops of these waves break and throw off a "spray" of saline water into the upper layer into which it mixes. There is no breaking at the bottom of the internal waves and therefore no spray of fresh water downward into the saline water.

For the *salt wedge estuary* (type D) the longitudinal section indicates the reason for its name. The saline water intrudes from the sea as a wedge below the river water. This situation is typical of rivers of large volume transport such as the Mississippi or the Fraser Rivers. It should be noted that as usual the section in Fig. 8.2 is exaggerated in the vertical direction; the salt wedge is really of a much smaller angle than shown in the figure so that the isohalines are, in fact, almost horizontal.

The salt wedge estuary has features in common with the stratified estuaries. There is a horizontal gradient of salinity at the bottom as in a slightly stratified estuary and a pronounced vertical salinity gradient as in a highly stratified estuary. The distinction is in the lack of saline water at the surface until it reaches the sea at the mouth of the estuary, because of the large river flow. One other feature of this type of estuary is that the salt wedge migrates up and down the estuary as the tide floods and ebbs, sometimes by several kilometres.

8.32 Estuarine circulation

A feature of the stratified estuaries is that the depth of the halocline, i.e. the thickness of the upper, low salinity layer, remains substantially constant from head to mouth of an estuary for a given river runoff. If the estuary width does not change much, the constancy of depth of the upper layer means that the cross-sectional area of the upper layer outflow remains the same while its volume transport increases because of the entrainment of salt water from below. In consequence the speed of the outflowing surface layer increases markedly along the estuary from head to moth. The increase in volume and speed can be very considerable, the outflow at the mouth being as much as 10 to 30 times the volume flow of the river. In his classical study of Alberni Inlet, a typical highly-stratified fjord-type estuary in British Columbia, Tully (1949) demonstrated the above features. He also showed that the depth of the upper layer decreased as the river runoff increased up to a critical value and thereafter increased as runoff increased.

The circulation described, with outflow of the upper layer and inflow below it is referred to as an *estuarine circulation* and has to be considered when practical problems of disposing of industrial effluent are under consideration. The circulation is dependent on several factors, among them the sill depth, the river runoff and the character of the outside water density distribution. If the sill is so shallow that it penetrates into the low-salinity outflowing upper layer the full estuarine circulation cannot develop and the subsurface inflow of saline water does not occur regularly. In consequence the deep water is not exchanged regularly and tends to become stagnant. This situation occurs in some of the smaller Norwegian fjords but is by no means typical of deep basin estuaries. Most of the fjords in that country and almost all of those on the west coasts of North and South America and New Zealand have sills which are deeper than the upper layer. In consequence the estuarine circulation is developed sufficiently to effect continual renewal of the deep water so that stagnation does not occur (Pickard, 1961; Pickard and Stanton, 1980). The rate of renewal is proportional to the circulation which is itself proportional to the river runoff. Fjord estuaries with small river runoff show more evidence of limited circulation in the form of low oxygen values than do those with large runoff. The depth of the sill has little effect as long as it is greater than the depth of the low-salinity outflowing upper layer.

The other major factor influencing the exchange of the deep basin water is a seasonal variation in density structure of the outside sea-water. Although the downward mixing of fresh water in an estuary is small it does occur to some extent. In consequence the salinity, and therefore the density of the basin water, tends to decrease slowly. If a change then occurs in the outside water such that the density outside becomes greater than that inside at similar levels above the sill depth, there will be an inflow of water from the sea. The inflowing water is likely to sink, although not necessarily to the bottom, in the estuary basin and so displace upward and outward some of the previously resident water. In this way the basin water becomes refreshed. In deep-sill estuaries this refreshment may occur annually, but in shallow-sill estuaries it may occur only at intervals of many years and the disturbance to the biological regime may be cataclysmic on these occasions (by displacing upward into the biotic zone the low-oxygen water from the bottom). It is this type of basin-water replacement which has been well documented for some Norwegian fjords (with very shallow sills), but it should not be considered characteristic of all fjord estuaries.

The above remarks should be regarded only as a brief description of some of the salient characteristics of estuaries, and it must be realized that the property distributions shown in Fig. 8.2 are smoothed and schematic. Real distributions show fine and meso-scale structure and detailed features, some general and some local. In particular, because the density structure is determined largely by the salinity distribution, temperature maxima and minima are quite common in the water column. The discussion in detail of the circulation is a matter for the dynamical oceanographer and will not be considered here. The mechanics of the process of mixing between fresh and salt water, in which tidal movements appear to play a large part as well as the effects of internal waves, are in the same category and much research remains to be done in this field. Reviews of both the descriptions and the dynamics of estuaries will be found in the texts by Dyer and by Officer listed in the Suggestions for Further Reading. A recent summary of the development of estuarine circulation ideas and of investigations of circulation in a coastal region (the Middle Atlantic Bight and Gulf of Maine, from about 36° to 43° N) has been given by Beardsley and Boicourt (1981).

It should be pointed out that estuarine characteristics and processes are observed in ocean areas as well as by the coast. In the north-east Pacific and in the Bay of Bengal where there is considerable river runoff, the density of the upper layer is controlled by the salinity rather than by temperature as is usually the case in the open ocean. The upper, low-salinity layer of perhaps 100 m depth in the north-east Pacific is less dense than the deeper, more saline water and the stability in the halocline between them inhibits mixing. In consequence the summer input of heat is trapped in the surface layer and a marked seasonal thermocline develops as shown in Fig. 4.6. In the Arctic Sea, the formation of the subsurface Arctic Water has been explained (Chapter 7) as the result of circulation and mixing processes similar to those described above for a coastal estuary.

CHAPTER NINE

Some Directions for Future Work

IN THIS book, an endeavour has been made to describe the kind of physical information available about the oceans at the present time. It has been illustrated with descriptions of some of the features of the ocean waters and of their circulation, but the reader will realize that these are only samples. For more complete information it would be necessary to consult more extensive texts, reviews and journal articles, such as those listed in the Bibliography. To conclude this presentation it will be pertinent to assess the state of our present knowledge and to indicate some of the information which we still need.

For the upper layer of the ocean, it is safe to say that we are acquainted with the main features of the circulation and of the distribution of water properties, but there are few oceanographers who are satisfied with the extent of their knowledge of even limited areas.

One of the first features of the upper-water structure described in this book was the thermocline. This occurs over most of the ocean but the reason for its continued existence is by no means clear. In low latitudes there is a net annual input of heat through the surface and together with the mixing due to the wind this might be expected to produce a steady deepening of the thermocline. The evidence available over the past 100 years or so is that this is not occurring. As an explanation for this Stommel (1958) suggested that the downward mixing of heat must be balanced by a net upward flow of cool deep water displaced by winter cooled water sinking in the North and South Atlantic. This is a very reasonable suggestion for the basic mechanism but it still leaves to be explained the differences in depth and in temperature gradient in the thermocline zone between different regions. Unfortunately the rate of upward flow is estimated to be so small as to be unmeasurable with present instruments (it is of the order of 1 cm/day) and therefore we can only infer it from heat budget calculations. These are only very approximate, since our techniques for measuring the component terms in the heat budget leave much to be desired in fundamental soundness, accuracy and convenience in use. In fact, one of the major problems in physical oceanography is to accurately resolve the terms in the heat budget. It is well known that the ocean stores heat energy which it exchanges with the atmosphere; these exchanges must have an impact on the earth's climate but our imperfect knowledge of how the ocean stores, transports and exchanges this heat prevents us from developing reliable prediction methods for climatic change. It should be noted that Iselin's suggestion (Chapter 7) for the mode of formation of Central Water Masses also provides a mechanism for the formation of the thermocline.

Many of the problems of the upper layers are to understand how oceanic motions are driven by the atmosphere; these are mainly problems in dynamic oceanography which will not be discussed here explicitly. To solve these problems requires quantitative observation of the behaviour of the surface currents. For example, the main character of the equatorial circulation has been well established but we have very few measurements of the volume transports of the currents and fewer of their variation of speed or position with time, or their rate of response to changes in winds. Recent transequatorial sections of water properties and current profiles (NORPAX Tahiti Shuttle Experiment) should improve our knowledge of this system. It is not out of the question that other currents remain to be discovered. For instance, the Pacific Equatorial Undercurrent which lies only 50 m below the surface and has at least half the transport of the Gulf Stream was not discovered until 1952, and we do not yet know in detail what happens to this major current at its eastern end. The Pacific South Equatorial Countercurrent was only recognized in 1959, and is still not well documented, nor are the eastward currents north of the North Equatorial Current in the western Pacific. The South Pacific is still imperfectly known, particularly the eastern half including the details of the very important *El Niño* phenomenon, and the currents on the east and west of Australia remain to be fully described.

Upwelling of subsurface water is important as it often brings to the surface the nutrients which promote the growth of plankton and hence of fish, while the strip of cool water along the coast affects the climate there and over the neighbouring land. Many of the regions of upwelling are known, but there may be others which are at present unrecognized. In the exchange of water within the upper layer, downwelling in regions of convergence is also important. Such regions are less easy to locate than upwelling regions but are still important in the circulation of the upper layer. Our great lack of information here is of the rate of vertical motion in both processes, and there does not seem to be in view at present any practical direct method to measure it in the upper layers. However, it is hoped that the Coastal Upwelling Ecosystems projects will yet yield a great deal of information about this process.

In general our picture of the upper ocean has evolved to the point where we understand the basic features and can describe the large-scale distribution of these features. As our ability to measure the ocean improves we continue to realize that the ocean is not as stationary as earlier assumed. Although many features fit well into our large-scale pattern, new scales of variability in both space and time have emerged. It is the task of the modern descriptive oceanographer to improve his observations of these scales of variability in a manner in which new dynamical interpretations will then be possible. This observational task requires new skills and approaches, especially as we realize the need for improved spatial sampling. Measurement techniques have been developed to sample intensively and continuously in time at a single location (e.g. a mooring) but it becomes quite costly to extend such measurements to a wide area. Recent improvements in infra-red imagery by satellite (Pl. 11) have revealed the complexity of the temperature distribution at the sea surface and suggested the need for similar synoptic sampling of the deeper ocean.

For the deep water we are even more poorly informed than for the upper water. For the Atlantic Ocean we do have a good idea of the general direction of movement of the main water masses, but have a much poorer knowledge of even the direction of flow of the deep waters of the Pacific and Indian Oceans. Stommel (1958, and with Arons, 1960a, 1960b) put forward a hypothesis, which has since been elaborated by others, that in the deep water there is a strong flow along the west sides of the oceans, with branches eastward and poleward to supply the upward flow needed to maintain the thermocline. This west-side flow is mainly southward from the Labrador Sea in the Atlantic and northward from the Southern Ocean in the Pacific and Indian Oceans. With regard to the speed of motion of the deep water we have little information for any ocean. Estimates of the age of the Atlantic subsurface water masses, from ^{14}C information have decreased from 700 years in 1960 to only 100 years in 1979, as our knowledge of the ocean structure and motion has improved. Even the age gives no more than an indication of the average speed. The measurements with Swallow floats, limited as they are, demonstrate that the instantaneous speeds are very variable. As stated before, the classical concept of a sluggish flow of deep water is certainly unrealistic and it is more likely that jets or filaments of fast-moving water are a basic feature of the deep circulation.

Another way of saying that we do not know much of even the mean speed of the deep water is to say that we have little idea of the rate of formation of water masses. For instance, is the North Atlantic Deep Water produced each year in small quantities (from the Norwegian Sea) or is it formed cataclysmically at irregular intervals as Worthington has suggested? Or do both processes play their part and, if so, to what extent?

Also, although they are perhaps of more concern to the dynamic than to the descriptive oceanographer, the mesoscale eddy features of the open ocean are under very active investigation at present and the significance of the redistribution of water characteristics and biota by Gulf Stream rings and possible similar phenomena near other major currents is being studied. Our understanding of these mesoscale eddies and their role in the general circulation is rapidly evolving. Eddies spawned by meanders of major currents such as the Gulf Stream, the Kuroshio and the Antarctic Circumpolar Current have been observed in formation and recognized as a means for removing energy from the mean flow. The generation of other, frequently smaller scale, eddies is a topic of current study. At present we have only a very limited knowledge of where these eddies generally occur and with what frequency. Efforts, such as POLYMODE, designed to study these features have clearly revealed that some regions may not be frequently populated by strong mesoscale eddies.

The T–S diagram has been described with its use for identifying subsurface water masses. These are believed to be formed at the surface, often in winter, as a result of heat and water exchange with the atmosphere. Over the period of 100 years for which oceanographic data suitable for comparison are available there have been marked variations in climatic conditions. It is remarkable that despite this the various water masses in the ocean have retained their oceanographic characteristics virtually unchanged. The "18°water" in the Atlantic is an example of what is almost a water type which has maintained its characteristics since the time of the *Challenger* expedition in 1873. It is clear that our knowledge of the heat budget requires considerable refinement in terms of input/output through the sea surface, of advection by ocean currents, and of the balance between advection and diffusion in the vertical direction. It is hoped that the data from the *Seasat* type of satellites and other ocean monitoring efforts will provide us with continuing large-scale coverage of the oceans to help us to understand both the overall picture and the seasonal variations for the ocean as well as providing information to help to understand the effect of ocean temperature anomalies on the atmosphere and our climate.

The physical properties of sea-ice and the reasons for its movement and distribution are neither fully described nor understood. To nations in the higher northern latitudes, Canada, Scandinavia and Russia, the ability to move shipping in the North Atlantic, Arctic and adjacent waters is very important. Although significant advances have been made in forecasting ice conditions, this cannot be done with certainty or very far ahead. The TIROS satellites from 1961 improved our observing capabilities immensely, even though cloud cover caused observing limitations. It is anticipated that the microwave and radar capabilities of future satellites will enable us to obtain much more complete information on ice distribution and of its changes with time.

In connection with the propagation of sound in the ocean, both for military purposes and for fish detection, there are still a vast number of data required on temperature distribution both on the macroscopic scale and also on the smaller scales associated with turbulence. It is also an interesting new area of research to try to relate the acoustic propagation characteristics to the structure of the water body through which the signal is propagating. Called "acoustic tomography", the programme of research hopes to scan the internal structure of the ocean in much the same manner that medical doctors now use to scan internal organs of the human body.

To advance his knowledge of the oceans the physical oceanographer needs more data in two forms – simultaneous coverage over large areas (together with meteorological data) and time-series observations over periods of at least a year in selected localities. It is quite clear that there are never going to be enough oceanographers and ships to satisfy the requirements and the only possible way in which these extra data are going to be obtained is by using satellites to observe as many as possible of the surface characteristics together with unmanned oceanographic observing buoys in large numbers for surface and subsurface characteristics. Some meteorological buoys are in use now and limited oceanographic measurements are being made from them but it will be some time before the flow of data from this source becomes significant.

For such buoys, the designer faces a number of difficulties. He is required to make an automatic device to measure temperature, salinity, etc., at a time when manually operated instruments for use from ships are still being developed. Buoys have to store large quantities of information, or preferably telemeter it to a base station frequently so that if a buoy is lost its store of information is not lost with it. There are also obvious problems in mooring such buoys so that they will withstand any weather. (A more serious problem is to obtain sufficient channels in the radiofrequency spectrum for telemetry.) Recent efforts have indicated that it may be more useful and practical to gather these data from freely drifting buoys which relay their data via satellite.

CTDs are in wide use today and are very helpful both in providing an immediate readout of water properties and in enabling data to be fed directly into a computer for correction and data assembly for reports and analysis. However, it is still considered good practice to check the CTD frequently against bottle-cast samples of both temperature and salinity to maintain a satisfactory level of absolute accuracy. Dissolved oxygen sensors are coming into use but have not yet reached the same level of acceptance as the temperature and salinity instruments. Immediate indication and recording as a function of depth permits the oceanographer to see on the spot what information he is getting, and to adjust his survey as necessary to make the best use of his ship time, and to follow particular features in space and time. Unfortunately, however, differences in instrument and calibration procedures have kept many of these CTD data from being submitted to the large data archives. As very few bottle casts are collected any more, our general ocean coverage is not increasing significantly. Combined with a general decrease in large-scale oceanographic surveys this lack of data submission has limited our historical data file on the ocean's density structure to measurements collected predominantly before 1970. Although the use of air-dropped XBTs (AXBTs) has made rapid upper ocean surveys possible, the potential of aircraft and satellites for use as platforms for oceanographic studies has still to be fully realized. The successful development of new instruments requires the combination of an assessment of the kind and accuracy of data required, the services of skilled instrument designers, and a full appreciation of the conditions under which the instruments have to be used at sea.

Finally, we draw attention to the dynamic oceanographer's technique for investigating the circulation by solving the equations of motion by numerical methods (*numerical modelling*) which is being increasingly used both for large- and for small-scale motion studies because the non-linear character of the equations makes it very difficult to obtain analytic solutions to them (i.e. solutions in the form of algebraic/trigonometric expressions for velocity and density distributions as a function of space and time). To use the modelling techniques the dynamic oceanographer needs from his descriptive colleagues the best available information on the density distribution (from temperature and salinity distributions), on air/sea exchange processes and on circulation, both to initiate his modelling studies and to test his results. Such procedures may be used to determine if the known ocean circulation develops when the modelling procedure is started from an ocean initially at rest, or it may be used to determine the effects on an existing circulation of changing the conditions or acting processes. Results of such studies may suggest ways of deploying our limited observational facilities to acquire information for more advanced models or theories, or may suggest means for solving practical problems associated with the sea. A short account of the techniques used in numerical modelling and of some of the results obtained is given in Pond and Pickard's *Introductory Dynamic Oceanography*.

The best account of the present state of physical oceanography is presented in *Evolution of Physical Oceanography*, a collection of scientific surveys in honour of Henry Stommel (see Bibliography).

It is clear that there are many intriguing problems yet to be solved in physical oceanography. There are extensive programmes of observation to be planned and instruments to be developed so that they may be carried out. The study and interpretation of the data and comparison with theoretical studies will yield a fuller understanding of the ocean circulations and, we can be sure, will reveal further problems to be investigated. Much of the research will be carried out by oceanographers who have no other motive than to improve their understanding of the behaviour of the oceans. The individual who likes to see some practical application for the new knowledge can be assured that it will at once be seized upon by those concerned with urgent practical questions in the development of fisheries, the solution of waste disposal problems and in coastal engineering.

Units used in Descriptive Physical Oceanography

Introduction

Formerly, physical oceanographers used a mixed system of (mostly) metric units but as the International System of Units (SI) is coming into general use it is used in this text. In SI there are *base units* and their multiples and *derived units*; a number of temporary units are accepted but will be phased out eventually. In the following sections, the base and derived units appropriate to our use will be listed and then these will be related to the older mixed units and to some specialized oceanographic units. The basic references used for SI practice are *The International System of Units (SI)*, reference CAN3–Z234.2–76, and *Canadian Metric Practice Guide*, reference CAN3–Z234.1–76, prepared by the Canadian Standards Association.

Relevant SI Units

Base Units

Quantity	Base unit	Abbreviation
Length	metre	m
Mass	kilogram	kg
Time	second	s
Thermodynamic temperature	kelvin	K

Multiples and Derived Units

Note that only units relevant to physical oceanography are given:

Quantity	Unit	Abbreviation or equivalent
Length	1 micrometre	$\mu m = 10^{-6}$ m
	1 centimetre	$cm = 10^{-2}$ m

	1 kilometre	km = 10^3 m
Volume	1 m^3	
	1 litre (L) = 1000 millilitres (mL) = 10^{-3} m^3	
	(the litre is not used for high-precision	
	measurements)	
Mass	1 gram	g = 10^{-3} kg
	1 tonne	t = 10^3 kg
Time	1 minute	min = 60 s
	1 hour	h = 60 min = 3600 s
	1 day (mean solar)	d = 24 h = 86,400 s
	1 year	yr (is not defined in SI,
		here taken as 365 d)
Frequency	1 hertz	Hz = 1 vibration/s
Speed	1 m/s = 100 cm/s	

(Note that the commonly used unit of 1 cm/s for current speeds is equivalent to 0.86 km/d or roughly 1 km/d or 300 km/yr.)

1 knot (kn) = 51 cm/s = 1.85 km/h

Density (ρ)   1 kg/m^3 = 10^3 g/m^3

Relative density (d) (formerly "specific gravity")—the value for liquids is given relative to pure water.

Force	1 newton	N = force required to give 1 kg
		mass an acceleration of 1 m/s^2.
Pressure (p)	1 pascal	Pa = 1 N/m^2
Energy	1 joule	J = 1 N.m
Power	1 watt	W = 1 J/s
Temperature (T)	The Celsius temperature (°C) is the difference between the	

thermodynamic temperature K and the temperature 273.15 K (unit = 1 K). (Temperature differences are stated in this text as K.)

Units used in Physical Oceanography and Some Numerical Examples

Volume transport   1 sverdrup   Sv = 10^6 m^3/s (Not approved by IAPSO but used in this text for conciseness.)

Salinity (S_A, S)   The definitions of absolute salinity (S_A) and of practical salinity (S) (PSS78) are given in Chapter 3. Expressed in this text as $\%_{oo}$ but "ppt" or "$\times 10^{-3}$" are also used.

Density (see-water (ρ) is a function of salinity, temperature and pressure, i.e. $\rho = \rho_{S,T,P}$. For sea-water of $S = 35\%_{oo}$, $T = 10.00°$C at standard atmospheric pressure $p = 101.325$ kPa (i.e. at zero hydrostatic pressure), the value of $\rho_{35,10,0} = 1026.97$ kg/m^3. Strictly speaking the measured oceanographic data for sea-water are for relative density (to pure water) which is dimensionless, but in physical equations it must be treated as density (dimensions = kg/m^3).

Sigma-t (σ_t)   This is introduced for convenience (Chapter 3). For the above sample of sea-water $\sigma_t = 26.97$ (it is usual to omit the units (kg/m^3) when stating the values for σ_t, which is usually used for descriptive purposes.

Specific volume ($\alpha = 1/\rho$) Unit = $1 \ m^3/kg = 10^3 \ cm^3/g$. For the above sea-water sample, $\alpha_{35,10,0} = 0.973 \ 74 \times 10^{-3} \ m^3/kg$. Also, $\alpha_{35,0,0} = 0.972 \ 64 \times 10^{-3} \ m^3/kg$.

Specific volume anomaly (δ) The SI unit is m^3/kg, the mixed unit is cm^3/g.

Thermosteric anomaly ($\Delta_{S,T}$) $\Delta_{S,T} = [(1000/\rho) - 0.972 \ 64] \times 10^{-3} \ m^3/kg$ so that the value of $\Delta_{S,T}$ for the above sea-water sample is $109.83 \times 10^{-8} \ m^3/kg$. In the mixed units system, the corresponding value is $109.83 \times 10^{-5} \ cm^3/g$ which is often stated as 109.83 centilitres/tonne (cL/t) to avoid writing the power of ten (1 cL/t = $10^{-8} \ m^3/kg$). In some texts, δ_t or Δ_T is used for the thermosteric anomaly.

Pressure   The appropriate SI unit for oceanography is $1 \ kPa = 10^3 \ Pa$. The previous basic unit was 1 bar, approximately equal to standard atmospheric pressure, and a unit used in oceanography was 1 decibar (dbar), equal to 10 kPa, which is about the pressure due to 1 m depth of sea-water. (The bar, dbar are not part of the SI.) The pressure in the open ocean at a depth of 1000 m is about 10,100 kPa (= 1010 dbar). Standard atmospheric pressure = 101.325 kPa (= 1.013 25 bar = 1013.25 mb = 760 mm Hg).

Heat energy   The appropriate SI unit is 1 joule (J) = 0.239 calorie using the conversion factor 1 calorie = 4.185 J.

Heat flow   The SI unit is 1 J/s = 1 watt (W).

Heat-flow density   The SI unit is $1 \ W/m^2 [= 2.39 \times 10^{-5} \ cal/(s \ cm^2)]$. A related mixed system unit was 1 langley (Ly) = $1 \ cal/cm^2$ and radiation heat flow was expressed in units of $1 \ Ly/min = 698 \ W/m^2$ or $1 \ Ly/d = 0.484 \ W/m^2$.

Latent heat   The SI unit is J/kg or kJ/kg. 1 cal/g = 4.185 kJ/kg so that for pure water, the latent heat of melting of 80 cal/g = 335 kJ/kg and the latent heat of evaporation at 10°C of 591 cal/g = 2.47×10^3 kJ/kg.

Electrical power   The SI unit = 1 kW = 1.34 h.p.

Electrical conductance   The SI unit is 1 siemens/m (S/m) = 1 mho/m = 0.01 mho/cm.

Dissolved oxygen   Usually quoted in units of 1 mL/L (= 1.43 mg/L = 0.089 mg at/L).

The appropriate SI unit for dissolved oxygen is micromole/kg ($\mu mol/kg$) and 1 mL/L = 43.3 $\mu mol/kg$ (for mean sea-water of $S = 34.7 \permil$, $T = 3.5\,°C$, $\sigma_t = 27.96$).

Bibliography

Two lists are given. The first is a general list of textbooks and sources of further information while the second is a list of specific journal articles or more specialized texts referred to in the body of our text.

Suggestions for Further Reading

The following texts may be consulted for more detailed information on some of the topics discussed in the present book:

BUDYKO, M. I.; *Climate and Life*, Academic Press, 1974, p. 508. A broad review of the climate of the earth with discussion of the relations between it and life. (English translation of the 1971 Russian text.)

DEFANT, A.; *Physical Oceanography*, Pergamon Press, 1960, p. 1319. An advanced level text. Volume 1, Pt. 1 is descriptive while Pt. 2 and all of Vol. 11 are dynamical.

DIETRICH, G. and K. KALLE; *General Oceanography—an Introduction*, Wiley, 1963, p. 588. A fairly comprehensive text on descriptive and dynamic oceanography with a little geology and biochemistry.

DYER, K. R.; *Estuaries—a Physical Introduction*, Wiley, 1973, p. 140. A summary of the descriptive and dynamic oceanography of estuaries.

GROSS, M. GRANT; *Oceanography—a View of the Earth*, Prentice-Hall, 2nd edn., 1977, p. 497. A survey of most aspects of oceanography, physical, geological, chemical and biological.

HARVEY, J. G.; *Atmosphere and Ocean; our Fluid Environments*, Artemis Press, 1976, p. 143. A coherent description of the characteristics and dynamics of the atmosphere and ocean and of their interaction, with simple mathematical developments.

HILL, M. N. (Ed.); *The Sea: Ideas and Observations*, Interscience, Vol. 1, 1962, p. 864. Physical oceanography. A collection of advanced papers on dynamical oceanography and energy transmission. Vol. 11, 1963, p. 554. Composition of sea-water. Comparative and descriptive oceanography. Further advanced papers on chemical synoptic and biological oceanography.

JERLOV, N. G.; *Marine Optics*, Elsevier, 1976, p. 231. An up-date of his earlier *Optical Oceanography* (1968) being a review of what is known about light in the sea.

KRAUS, E. G.; *Atmosphere—Ocean Interaction*, Clarendon Press, 1972, p. 271. A fairly advanced discussion of radiation exchange at the surface of the ocean and of the exchange of energy and momentum between the atmosphere and ocean.

LAUFF, G. H. (Ed.); *Estuaries*, Amer. Assn. Adv. Sci., Publ. No. 83, p. 757. A collection of papers on a variety of aspects of the oceanography of estuaries.

MCLELLAN, H. J.; *Elements of Physical Oceanography*, Pergamon, 1965, p. 150. An introduction to descriptive and dynamic oceanography.

NEUMANN, G. and W. J. PIERSON; *Principles of Physical Oceanography*, Prentice-Hall, 1966, p. 545. Moderately advanced text on descriptive and dynamic oceanography.

OFFICER, C. B.; *Physical Oceanography of Estuaries and Associated Coastal Waters*, Wiley, 1976, p. 465. A moderately advanced account of the description and dynamic theory with applications to typical areas around the world.

PERRY, A. H. and J. M. WALKER; *The Ocean–Atmosphere System*, Longman, 1977, p. 160. A good account of the ocean–atmosphere system with many illustrations and references to original literature.

POND, S. and G. L. PICKARD; *Introduction to Dynamic Oceanography*, Pergamon, 1978, p. 258. An introduction to the basic principles and equations of dynamic oceanography, the role of the non-linear terms, currents without and with friction, numerical modelling, waves and tides.

SHEPARD, F. P.; *Submarine Geology*, Harper & Row, 1973, p. 517. A description of the sediments and structure of the ocean basins and of their origins.

SHEPARD, F. P.; *Geological Oceanography*, Crane, Russak, 1977, p. 214. A well-illustrated account of many recent developments in this field.

STOMMEL, H.; *The Gulf Stream*, University of California Press, 2nd edn., 1965, p. 248. Both a description of this ocean feature and an excellent introduction to physical oceanography for upper-year undergraduates and graduate students in physics.

SVERDRUP, H. U., M. W. JOHNSON and R. H. FLEMING; *The Oceans, their Physics, Chemistry and General Biology*, Prentice-Hall, New York, 1946, p. 1087. A comprehensive reference book on all aspects of oceanography to about 1942.

TCHERNIA, P.; *Descriptive Regional Oceanography*, Pergamon, 1978, p. 250. Describes the physical oceanography of the oceans and seas in more detail than the present text with numerous illustrations and fold-out charts.

VAN DORN, W. G.; *Oceanography and Seamanship*, Dodd Mead, 1974, p. 481. An interesting and simply written account of how knowledge of the ocean is applied to practical seamanship.

VON ARX, W. S.; *An Introduction to Physical Oceanography*, Addison Wesley, 1962, p. 422. A stimulating introduction to many aspects of physical oceanography, with special emphasis on current measurement and the use of laboratory scale-models. Also includes study questions and a chronological list of significant events in the marine sciences.

WARREN, B. A. and C. WUNSCH (Eds.); *Evolution of Physical Oceanography*, MIT Press, 1981, p. 623. An excellent series of survey articles on the general ocean circulation, physical processes, techniques and ocean/atmosphere interaction tracing the development of these fields and laying out the state of our knowledge now. Prepared in honour of Henry Stommel. Advanced but essential to the library of any serious physical oceanographer.

WEYL, P. K.; *Oceanography—an Introduction to the Marine Environment*, Wiley, 1970, p. 535. Mostly physical, chemical and geological with some biology and ecology.

Oceanography—Readings from Scientific American; Freeman, 1971, p. 417. A collection of stimulating articles on many aspects of oceanography.

Ocean Science—Readings from Scientific American; Freeman, 1977, p. 307. Contains articles additional to those in *Oceanography*, 1971.

Four accounts of the development of oceanography are:

DEACON, M. B.; *Scientists and the Sea 1650–1900: a Study of Marine Science*, Academic Press, 1971, p. 398.

RAITT, H. and B. MOULTON; *Scripps Institution of Oceanography: the First Fifty Years*, Ward Ritchie, 1967, p. 217.

SCHLEE, S.; *The Edge of an Unfamiliar World: a History of Oceanography*, Dutton, 1973, p. 397.

SHOR, E. N.; *Scripps Institution of Oceanography: Probing the Oceans 1936–1976*, p. 502.

Some information on the properties of sea-water and on methods will be found in the following publications:

IVANOFF, A.; *Introduction à l'Océanographie, Propriétés Physiques et Chimiques des Eaux de Mer*, Vol. 1, Vuibert, Paris, 1972, p. 206. A compact summary of these properties and of distributions in the oceans. Volume 2 on the heat budget and on electrical and acoustical properties is planned.

RILEY, J. P. and G. SKIRROW (Eds.); *Chemical Oceanography*, Academic Press, 1965, Vol. 1, p. 712. Includes a chapter with references on the physical properties of sea-water.

Handbook of Oceanographic Tables; U. S. Naval Oceanographic Office, Special Publication 68, Washington, DC, 1966, p. 1966, p. 427. A collection of tables of use to oceanographers.

International Oceanographic Tables, Vol. 1; National Institute of Oceanography of Great Britain, and Unesco, 1966, p. 128. For converting conductivity ratio to salinity of sea-water. (To be replaced soon by tables based on the PSS78.)

LAFOND, E. C.; *Processing Oceanographic Data*, U. S. Naval Oceanographic Office Publication 614, Washington, DC, 1951, p. 114. A compilation of tables needed for correcting thermometers, calculating density and specific volume, etc.

Instruction Manual for obtaining Oceanographic Data; U. S. Naval Oceanographic Office, Washington, DC, Publ. 607., 3rd edn. 1968, Reprint 1970, p. 210. A description of routine oceanographic procedures at sea (bottle sampling, mechanical BT, coring, etc.) and of the older standard instruments.

The classical papers in oceanography are scattered through many scientific journals and reports of expeditions. Many of the recent papers may be found in:

Deep-Sea Research, Pergamon Press, Oxford (since 1953).

Estuarine and Coastal Marine Science, Academic Press, London (since 1973).

Journal du Conseil (from 1926) and *Annales Biologiques* (from 1939), Conseil Perm. Intern, pour l'Explor. de la Mer, Copenhagen.

Journal of Fisheries and Aquatic Science, formerly *Journal of the Fisheries Research Board of Canada*, Ottawa (since 1934).

Journal of Geophysical Research, Amer. Geophys. Union, Washington, DC (since 1959).

Journal of Marine Research, Sears Foundation for Marine Research, New Haven, Connecticut (since 1939).

Journal of Physical Oceanography, Amer. Met. Society, Lancaster, Pa. (since 1971).

Limnology and Oceanography, Amer. Soc. of Limnol. and Oceanogr., Lawrence, Kansas (since 1956).

Two annual reviews of aspects of oceanography are:

Oceanography and Marine Biology, M. BARNES (Ed.), Aberdeen University Press (from 1963).

Progress in Oceanography, M. V. ANGEL and J. O'BRIEN (Eds.), Pergamon Press (from 1964).

An interesting quarterly with simply written but authoritative articles on many aspects of oceanography is:

Oceanus, Woods Hole Oceanographic Institution, Woods Hole, Mass. (since 1952).

References to Journal and Review Articles

The locations for the references (by author and date) made in the text are presented here for the reader who wishes to pursue a topic further. A few of the references are to articles of historic interest, e.g. first announcements of newly recognized ocean features or developed techniques, others are to review articles or to those having more detail than can be included in an introductory textbook. We make no claim that this is a list of the "170 most important articles" in descriptive oceanography. The reader should note that two useful sources for references are *The Oceans* by Sverdrup, Johnson and Fleming (1946) for material published before about 1942 and *Evolution of Physical Oceanography* edited by Warren and Wunsch (1981) for later material.

AAGAARD, K. and P. GRIESMAN, (1975) Toward new mass and heat budgets for the Arctic Ocean. *Journal of Geophysical Research*, **80**, 3821–3827.

ÅNGSTRÖM, A. (1920) Application of heat radiation measurements to the problems of the evaporation from lakes and the heat convection at their surfaces. *Geografische Annalen*, **3**, p. 16.

BAILEY, W. B. (1957) Oceanographic features of the Canadian Archipelago. *Journal, Fisheries Research Board of Canada*, **14**, 731–769.

BAINBRIDGE, A. E. (1976) *GEOSECS Atlantic Expedition*, Vol. 2 *Sections and Profiles*, National Science Foundation, Wash., DC, p. 198.

BAKER, D. J. (1981) Ocean instruments and experiment design. Ch. 14, pp. 396–433, in *Evolution of Physical Oceanography*, B. A. WARREN and C. WUNSCH (Eds.), MIT Press.

BARBER, R. T. (1877) The JOINT-1 expedition of the Coastal Upwelling Ecosystems Analysis programme. *Deep-Sea Research*, **24**, 1–6.

BARNETT, T. P. (1977) An attempt to verify some theories of El Niño. *Journal of Physical Oceanography*, **7**, 633–647.

BEARDSLEY, R. C. and W. C. BOICOURT (1981) On estuarine and continental-shelf circulation in the Middle Atlantic Bight. Ch. 7, pp. 198–233 in *Evolution of Physical Oceanography*, B. A. WARREN and C. WUNSCH (Eds.), MIT Press.

BJERKNES, V., J. BJERKNES, H. SOLBERG and T. BERGERON (1933) *Physikalische Hydrodynamik*. Springer-Verlag, p. 797.

BJERKNES, V. and J. W. SANDSTRÖM (1910) *Dynamic Meteorology and Hydrography, Part 1, Statics.* Carnegie Institute, Washington, Publ. No. 88, p. 146.

BOWEN, I. S. (1926) The ratio of heat losses by conduction and by evaporation from any water surface. *Physical Review,* **27**, 779–787.

BROECKER, W. S. (1979) A revised estimate for the radiocarbon age of the North Atlantic Deep Water. *Journal of Geophysical Research,* **84**, 3218–3226.

BROWN, N. L. and B. V. HAMON (1961) An inductive salinometer. *Deep-Sea Research,* **8**, 65–75.

BRYDEN, H. L. (1978) Mean upwelling velocities on the Oregon continental shelf during summer 1973. *Estuarine and Coastal Marine Science,* **7**, 311–327.

BRYDEN, H. L. and R. D. PILSSBURY (1977) Variability of deep flow in the Drake Passage from year-long current measurements. *Journal of Physical Oceanography,* **7**, 803–810.

CAMERON, W. M. and D. W. PRITCHARD (1963) Estuaries. Ch. 15, pp. 306–324, in *The Sea: Ideas and Observations,* M. N. HILL (Ed.), Vol. 2, Wiley-Interscience.

CARMACK, E. C. and K. AAGAARD (1973) On the deep water of the Greenland Sea. *Deep-Sea Research,* **20**, 687–715.

CARRITT, D. E. and J. H. CARPENTER (1958) The composition of sea-water and the salinity–chlorinity–density problem. pp. 67–86 in *Physical and Chemical Properties of Sea-water,* National Academy of Science-National Research Council, Publ. 600.

CHENEY, R. C. and J. G. MARSH (1981) *Seasat* altimeter observations of dynamic topography in the Gulf Stream. *Journal of Geophysical Research,* **86**, 473–484.

CLARKE, R. A., H. HILL, R. F. REINIGER and B. A. WARREN (1980) Current system south and east of the Grand Banks of Newfoundland. *Journal of Physical Oceanography,* **10**, 25–65.

COACHMAN, L. K. and K. AAGAARD (1974) Physical oceanography of Arctic and Subarctic seas. Ch. 1, pp. 1–72 in *Marine Geology and Oceanography of the Arctic Seas,* Y. HERMANN (Ed.), Springer-Verlag.

COCHRANE, J. D. (1958) The frequency distribution of water characteristics in the Pacific Ocean. *Deep-Sea Research,* **5**, 111–127.

COOPER, L. H. N. (1955) Deep water movements in the North Atlantic as a link between climatic changes around Iceland and biological productivity of the English Channel and Celtic Sea. *Journal of Marine Research,* **14**, 347–362.

COX, R. A., M. J. McCARTNEY and F. CULKIN (1970) The specific gravity/salinity/temperature relationship in natural sea-water. *Deep-Sea Research,* **17**, 679–689.

COX, R. A. and N. D. SMITH (1959) The specific heat of sea-water. *Proceedings, Royal Society of London,* Series A, **252**, 51–62.

CREASE, J. (1965) The flow of Norwegian Sea water through the Faroe Bank Channel. *Deep-Sea Research,* **12**, 143–150.

CROMWELL, T., R. B. MONTGOMERY and E. D. STROUP (1954) Equatorial undercurrent in Pacific Ocean revealed by new methods. *Science,* **119**, 648–649.

DAUPHINEE, T. M. and H. P. KLEIN (1977) The effect of temperature on the electrical conductivity of sea-water. *Deep-Sea Research,* **24**, 891–902.

DEACON, G. E. R. (1937) The hydrology of the Southern Ocean. *Discovery Reports,* **15**, 1–124.

DEGENS, E. T. and D. A. ROSS (Eds.) (1969) *Hot Brines and Recent Heavy Metal Deposits in the Red Sea.* Springer-Verlag, p. 600.

DIETRICH, G. (1969) Atlas of the hydrography of the northern North Atlantic Ocean based on the Polar Front Survey of the International Geophysical year, winter and summer 1958. *Conseil International pour l'Exploration de la Mer,* p. 140.

DODIMEAD, A. J., F. FAVORITE and T. HIRANO (1963) Review of oceanography of the Subarctic Pacific region. Part 2, pp. 1–195, of *Salmon of the Pacific Ocean,* Bulletin 13, International North Pacific Fisheries Commission.

DONGUY, J. and C. HENIN (1975) Evidence of the South Tropical Countercurrent in the Coral Sea. *Australian Journal of Marine and Freshwater Research,* **26**, 405–409.

DONN, W. L. and D. SHAW (1966) The heat budgets of an ice-free and an ice-covered Arctic Ocean. *Journal of Geophysical Research,* **71**, 1087–1093.

DORONIN, Y. P. and D. E. KHESIN (1975) *Sea Ice.* Amerind Publishing Company (Trans., 1977), p. 323.

DÜING, W. and D. JOHNSON (1972) High resolution current profiling in the Straits of Florida. *Deep-Sea Research,* **19**, 259–274.

EKMAN, V. W. (1905) On the influence of the earth's rotation on ocean currents. *Royal Swedish Academy of Science, Arkiv för matematik, astronomi och fysik*, **2**, No. 11, 1–53.

EKMAN, V. W. (1908) Die Zusammendrückbarkeit des Meereswassers. *Conseil Permanent International pour l'Exploration de la Mer, Publications de Circonstance*, No. 43, p. 47.

EKMAN, V. W. (1953) Studies on ocean currents. Results of a cruise on board the *Armauer Hansen* in 1930, Parts I and II. *Geofysiske Publicasjoner*, **19**, p. 106 and 122.

EMERY, W. J. (1977) Antarctic Polar Frontal Zone from Australia to the Drake Passage. *Journal of Physical Oceanography*, **7**, 811–822.

EMERY, W. J. (1980) The *Meteor* Expedition, an ocean survey. pp. 690–702 in *Oceanography, the Past*, M. SEARS and D. MERRIMAN (Eds.), Springer-Verlag.

EMERY, W. J. and J. S. DEWAR (1982) Mean temperature–salinity, salinity–depth and temperature–depth curves for the North Atlantic and the North Pacific. *Progress in Oceanography*, Pergamon (in press).

ESTERSON, G. L. (1957) Induction conductivity indicator. A new method for conductivity measurement at sea. *Chespeake Bay Institute, Technical Report* No. 14, p. 183.

FARADAY, M. (1832) Bakerian Lecture—Experimental researches in electricity. *Philosophical Transactions*, Royal Society of London, Part 1, 163–177.

FAVORITE, F., A. J. DODIMEAD and K. NASU (1976) *Oceanography of the Subarctic Pacific Region, 1960–71*. Bulletin 33, International North Pacific Fisheries Commission, p. 187.

FEDEROV, K. N. (1976) *The Thermohaline Finestructure of the Ocean* (Trans. D. A. BROWN), Pergamon, p. 170.

FIRING, E. (1981) Current profiling in the NORPAX Tahiti Shuttle. *Tropical Ocean-atmosphere Newsletter*, No. 5, January 1981, pp. 1, 8, 9. (Unpublished manuscript.)

FLEMING, R. H. (1939) Tables for Sigma-T. *Journal of Marine Research*, **2**, 9–11.

FOFONOFF, N. P. (1981) The Gulf Stream System. Ch. 4, pp. 112–139, in *Evolution of Physical Oceanography*, B. A. WARREN and C. WUNSCH (Eds.), MIT Press.

FORCHHAMMER, G. (1865) On the composition of sea-water in the different parts of the ocean. *Philosophical Transactions*, Royal Society of London, **155**, 203–262.

FUGILISTER, F. C. (1955) Alternative analyses of current surveys. *Deep-Sea Research*, **2**, 213–229.

FUGILISTER, F. C. (1960) *Atlantic Ocean Atlas of Temperature and Salinity Profiles and Data from the I.G.Y. of 1957–58*. Woods Hole Oceanographic Institution Atlas Series No. 1, p. 209.

FUGILISTER, F. C. and L. V. WORTHINGTON (1951) Some results of a multiple-ship survey of the Gulf Stream. *Tellus*, **3**, 1–14.

GERSTNER, F. (1802) *Theorie der Wellen u.s.w.* Abh. Kgl. Böhm. Ges. Wiss. Prague.

GILL, A. E. (1973) Circulation and bottom water production in the Weddell Sea. *Deep-Sea Research*, **20**, 111–140.

GUELKE, R. W. and C. A. SCHONTE-VANNECK (1947) The measurement of sea-water velocities by electromagnetic induction. *Journal of the Institute of Electrical Engineering*, **94**, 71–74.

HAMON, B. V. (1955) A temperature–salinity–depth recorder. *Conseil Permanent International pour l'Exploration de la Mer, Journal du Conseil*, **21**, 22–73.

HAMON, B. V. and N. L. BROWN (1958) A temperature–chlorinity–depth recorder for use at sea. *Journal of Scientific Instruments*, **35**, 452–458.

HAMON, B. V. and T. J. GOLDING (1980) Physical oceanography of the Australian region. Commonwealth Scientific and Industrial Research Organisation, Fisheries and Oceanography Report 1977–79, 1–8.

HELLAND-HANSEN, B. (1916) Nogen hydrografiscke metodor. *Forh. Skandinaviske Naturforske möte*, **16**, 357–359.

HELLAND-HANSEN, B. (1934) The Sognefjord section. pp. 257–274 in the *James Johnstone Memorial Volume*, Liverpool University Press.

HISARD, Ph., J. MERLE and B. VOITURIEZ (1970) The equatorial undercurrent at 170° E in March and April, 1967. *Journal of Marine Research*, **28**, 281–303.

HISARD, Ph. and P. RUAL (1970) Courant équatorial intermédiaire de l'océan Pacifique et contrecourantes adjacents. *Cahiers O.R.S.T.O.M.*, Série océanographique, **8**, 21–45.

ISELIN, C.O'D. (1936) A study of the circulation of the western North Atlantic. *Papers in Physical Oceanography and Meteorology*, **4**, p. 101.

ISELIN, C.O'D. (1939) The influence of vertical and lateral turbulence on the characteristics of the waters at mid-depths. *Transactions, American Geophysical Union*, **20**, 414–417.

JARRIGE, F. (1973) Temperature inversions in the equatorial Pacific. pp. 47–53 in *Oceanography of the South Pacific 1972*, R. FRASER (Ed.), New Zealand National Commission for Unesco.

JENKINS, G. M. and D. G. WATTS (1969) *Spectral Analysis and its Applications*. Holden-Day, p. 525.

KNAUSS, J. A. (1960) Measurements of the Cromwell Current. *Deep-Sea Research*, **6**, 265–286.

KNAUSS, J. A. (1961) The structure of the Pacific Equatorial Countercurrent. *Journal of Geophysical Research*, **66**, 143–155.

KNUDSEN, M. (1900) Ein hydrographische Lehrsatz. *Annalen der Hydrographie und Marinen Meteorologie*, **28**, 316–320.

KNUDSEN, M. (Ed.) (1901) *Hydrographical Tables*. G.E.C. Gad, Copenhagen, p. 63.

KOSSINNA, E. (1921) Die Teifen des Weltmeeres. *Institute für Meereskunde,Veroff. Geogr. naturwiss.* **9**, 70.

KREMLING, K. (1972) Comparison of specific gravity in natural sea-water from hydrographical tables and measurements by a new density instrument. *Deep-Sea Research*, **19**, 377–383.

KULP, J. L., L. E. TRYON, W. R. ECKELMANN and W. A. SNELL (1952) Lamont natural radiocarbon measurements. *Science*, **116**, 409–414.

LAEVASTU, T. (1963) Energy exchange in the North Pacific; its relation to weather and its oceanographic consequences. *Hawaii Institute of Geophysics*, Report No. 29, p. 15.

LAPLACE, P. S. (1775) Recherches sur plusiers points du système du monde. *Mémoires de l'Académie Royale des Sciences, Paris*, **88**, 75–182.

LAZIER, J. R. N. (1973) The renewal of Labrador Sea water. *Deep-Sea Research*, **20**, 341–353.

LEE, A. and D. ELLETT (1965) On the contribution of overflow water from the Norwegian Sea to the hydrographic structure of the North Atlantic Ocean. *Deep-Sea Research*, **12**, 129–142.

LEETMA, A., J. P. McREARY and D. W. MOORE (1981) Equatorial currents; observations and theory. Ch. 6, pp. 184–196 in *Evolution of Physical Oceanography*, B. A. WARREN and C. WUNSCH (Eds.), MIT Press.

LEWIS, E. L. (1980) The Practical Salinity Scale 1978 and its antecedents. *IEEE Journal of Oceanic Engineering*, **OE-5**, 3–8.

LEWIS, E. L. and N. P. FOFONOFF (1979) A practical salinity scale. *Journal of Physical Oceanography*, **9**, 446.

LEWIS, E. L. and R. G. PERKIN (1978) Salinity: its definition and calculation. *Journal of Geophysical Research*, **83**, 466–478.

LONGUET-HIGGINS, M. S., M. E. STERN and H. STOMMEL (1954) The electric field induced by ocean currents and waves with applications to the method of towed electrodes. *Papers in Physical Oceanography and Meteorology*, **13**, No. 1, p. 37.

LYNN, R. J. and J. L. REID (1968) Characteristics and circulation of deep and abyssal waters. *Deep-Sea Research*, **15**, 577–598.

MALMGREN, F. (1927) On the properties of sea-ice. *Norwegian North Polar Expedition with the Maud, 1918–1925, Scientific Research*, **1**, No. 5, p. 67.

MAMEYEV, O. I. (1975) *Temperature–Salinity Analysis of World Ocean Waters*. Elsevier, p. 374.

MANN, C. R. (1967) The termination of the Gulf Stream and the beginning of the North Atlantic Current. *Deep-Sea Research*, **14**, 337–359.

MANTYLA, A. W. (1975) On the potential temperature in the abyssal Pacific Ocean. *Journal of Marine Research*, **33**, 341–354.

MAURY, M. F. (1874) *The Physical Geography of the Sea*. Nelson, p. 493.

MENARD, H. W. and S. M. SMITH (1966) Hypsometry of ocean basin provinces. *Journal of Geophysical Research*, **71**, 4305–4325.

MERZ, A. and G. WÜST, (1922) Die Atlantische Vertikalzirkulation. *Zeitschrift der Gesellschaft für Erdkunde zu Berlin*, Jahrgang 1922, pp. 1–35.

METCALF, W. G. (1960) A note on water movement in the Greenland–Norwegian Sea. *Deep-Sea Research*, **7**, 190–200.

METCALF, W. G., A. D. VOORHIS and N. C. STALCUP (1962) The Atlantic Equatorial Undercurrent. *Journal of Geophysical Research*, **67**, 2499–2508.

MILLERO, F. J. (1967) High precision magnetic float densimeter. *Review of Scientific Instruments*, **38**, 1441–1444.

MILLERO, F. J., C-T. CHEN, A. BRADSHAW and K. SCHLEICHER (1980) A new high pressure equation of state for sea-water. *Deep-Sea Research*, **27A**, 255–264.

MILLERO, F. J., A. GONZALEZ and G. K. WARD (1976) The density of sea-water solutions as a function of temperature and salinity. *Journal of Marine Research*, **34**, 61–93.

MILLERO, F. J., G. PERRON and J. E. DESNOYERS (1973) The heat capacity of sea-water solutions from 5 to 35° C and 0.5 to 22‰ chlorinity. *Journal of Geophysical Research*, **78**, 4499–4507.

MODE GROUP (1976) Ocean eddies. *Oceanus*, **19**, 1–86.

MODE GROUP (1978) The Mid-Ocean Dynamics Experiment. *Deep-Sea Research*, **25**, 859–910.

MONTGOMERY, R. B. (1938) Circulation in the upper layer of the southern North Atlantic deduced with the aid of isentropic analysis. *Papers in Physical Oceanography and Meteorology*, **6**, No. 2, p. 55.

MONTGOMERY, R. B. (1958) Water characteristic of Atlantic Ocean and of World Ocean. *Deep-Sea Research*, **5**, 134–148.

MONTGOMERY, R. B. (1962) Equatorial Undercurrent observations in review. *Journal of the Oceanographical Society of Japan*, 20th Anniversary Volume, 487–498.

MONTGOMERY, R. B. and WARREN S. WOOSTER (1954) Thermosteric anomaly and the analysis of serial oceanographic data. *Deep-Sea Research*, **2**, 63–70.

NAMIAS, J. (1972) Large-scale and long-term fluctuations in some atmospheric and oceanic variables. pp. 27–48 in Nobel Symposium 20: *The Changing Chemistry of the Oceans*, D. DRYSSEN and D. JAEGER (Eds.), Wiley-Interscience.

NAMIAS, J. (1975) Stabilization of atmospheric circulation patterns by sea-surface temperature. *Journal of Marine Research*, **33** (Supplement), 53–60.

NANSEN, F. (1912) Die Bodenwasser und die Abkülung des Meeres. *Internationale Revue der gesamten Hydrobiologie und Hydrographie*, **5**, No. 1, p. 42.

NEIMANN, A. C. and D. A. GILL (1961) Circulation of the Red Sea in early summer. *Deep-Sea Research*, **8**, 223–235.

NEWTON, I. (1687) *Philosophia naturalis principia mathematica*, London.

NOWLIN, W. D., T. WHITWORTH and R. D. PILLSBURY (1977) Structure and transport of the Antarctic Circumpolar Current from short-term measurements. *Journal of Physical Oceanography*, **7**, 787–802.

OORT, A. H. and T. H. VANDER HAAR (1976) On the observed annual cycle in the ocean–atmosphere heat balance over the northern hemisphere. *Journal of Physical Oceanography*, **6**, 781–800.

PARKER, C. E. (1971) Gulf Stream rings in the Sargasso Sea. *Deep-Sea Research*, **18**, 981–993.

PICKARD, G. L. (1961) Oceanographic features of inlets in the British Columbia mainland coast. *Journal, Fisheries Research Board of Canada*, **18**, 907–999.

PICKARD, G. L. (1977) A review of the physical oceanography of the Great Barrier Reef and Western Coral Sea. *Australian Institute of Marine Science*, Monograph No. 2, p. 134.

PICKARD, G. L. and B. R. STANTON (1980) Pacific fjords—a review of their water characteristics. pp. 1–51 in *Fjord Oceanography*, H. J. FREELAND, D. M. FARMER and C. D. LEVINGS (Eds.), Plenum Press.

POLLACK, M. J. (1958) Frequency distribution of potential temperature and salinities in the Indian Ocean. *Deep-Sea Research*, **5**, 128–133.

PRITCHARD, D. W. (1952) Estuarine hydrography. *Advances in Geophysics*, **1**, 243–280.

REID, J. L. (1959) Evidence of a South Equatorial Countercurrent in the Pacific Ocean. *Nature*, **184**, 209–210.

REID, J. L. (1965) Intermediate Waters of the Pacific Ocean. *Johns Hopkins Oceanographic Studies*, No. 2, p. 85.

REID, J. L. (1973) Northwest Pacific Ocean waters in winter. *Johns Hopkins Oceanographic Studies*, No. 5, p. 96.

REID, J. L. and R. J. LYNN (1971) On the influence of the Norwegian–Greenland and Weddell Seas upon the bottom waters of the Indian and Pacific Oceans. *Deep-Sea Research*, **18**, 1063–1088.

RICHARDSON, P. L. (1980) Gulf Stream ring trajectories. *Journal of Physical Oceanography*, **10**, 90–104.

RICHARDSON, P. L., R. E. CHENEY and L. V. WORTHINGTON (1978) A census of Gulf Stream rings, Spring 1975. *Journal of Geophysical Research*, **83**, 6136–6144.

RING GROUP (1981) Gulf Stream cold-core rings: their physics, chemistry and biology. *Science*, **212**, 1091–1100.

ROTSCHI, H. (1970) Variation of equatorial currents. pp. 75–83 in *Scientific Exploration of the South Pacific*, W. S. WOOSTER (Ed.), U.S. National Academy of Science.

ROTSCHI, H. (1973) Hydrology at 170° E in the South Pacific, pp. 113–128 in *Oceanography of the South Pacific 1972*, R. FRASER (Ed.), New Zealand National Commission for Unesco.

ROTSCHI, H. Ph. HISARD and F. JARRIGE (1972) Les Eaux du Pacifique Occidental à 170° E entre 20° S et 4° N. *Travaux et Documents de l'O.R.S.T.O.M.*, Paris, No. 19, p. 113.

SANKEY, T. (1973) The formation of deep water in the Northwestern Mediterranean. *Progress in Oceanography*, B. A. WARREN (Ed.), **6**, 159–179.

SMITH, E. H., F. M. SOULE and O. MOSBY (1937) *Marion and General Greene* expeditions to Davis Strait and Labrador Sea 1928–1935. Scientific results, part 2, physical oceanography. *U.S. Coast Guard Bulletin* No. 19, p. 259.

SMITH, R. L. (1968) Upwelling. *Oceanography and Marine Biology, Annual Reviews*, **6**, 11–46.

STOKES, G. (1847) On the theory of oscillatory waves. *Cambridge Transactions*, **8**, pp. 212.

STOMMEL, H. (1958) The abyssal circulation. *Deep-Sea Research*, **5**, 80–82.

STOMMEL, H. and A. B. ARONS (1960a) On the abyssal circulation of the World Ocean, I. Stationary planetary flow patterns on a sphere. *Deep-Sea Research*, **6**, 140–154.

STOMMEL, H. and A. B. ARONS (1960b) On the abyssal circulation of the World Ocean, II. An idealized model of the circulation pattern and amplitude in oceanic basins. *Deep-Sea Research*, **6**, 217–233.

STOMMEL, H., P. NIILER and D. ANATI (1978) Dynamic topography and recirculation of the North Atlantic. *Journal of Marine Research*, **36**, 449–468.

STOMMEL, H. and K. YOSHIDA (Eds.) (1972) *Kuroshio: Physical Aspects of the Japan Current*. University of Washington Press, p. 517.

STRICKLAND, J. D. H. and T. R. PARSONS (1972) *A Practical Handbook of Sea-water Analysis*. Fisheries Research Board of Canada, Bulletin 167 (Second Edition), p. 310.

SWALLOW, J. C. (1955) A neutral-buoyancy float for measuring deep currents. *Deep-Sea Research*, **3**, 74–81.

SWALLOW, J. C. and L. V. WORTHINGTON (1957) Measurements of deep currents in the western North Atlantic. *Nature*, **179**, 1183–1184.

SWALLOW, J. C. and L. V. WORTHINGTON (1961) An observation of a deep counter-current in the western North Atlantic. *Deep-Sea Research*, **8**, 1–19.

TAIT, R. I. and M. R. HOWE (1968) Some observations of thermohaline stratification in the deep ocean. *Deep-Sea Research*, **15**, 275–280.

TAIT, R. I. and M. R. HOWE (1971) Thermohaline staircase. *Nature*, **231**, 178–179.

THOULET, J. and A. CHEVALLIER (1889) Sur la chaleur spécifique de l'eau de mer à divers degrés de dilution et de concentration. *Comptes Rendues de l'Académie des Sciences, Paris*, **108**, 794–796.

TULLY, J. P. (1949) Oceanography and prediction of pulp-mill pollution in Alberni Inlet. *Fisheries Research Board of Canada*, Bulletin 83, pp. 169.

TURNER, J. S. (1981) Small-scale mixing processes. Ch. 8, pp. 236–262, in *Evolution of Physical Oceanography*, B. A. WARREN and C. WUNSCH (Eds.), MIT, Press.

VAN RIEL, P. M. (1934) The bottom configuration in relation to the flow of bottom water. *Snellius* Expedition, Vol. 2, Ch. 2, p. 63.

WALLACE, W. J. (1974) *The Development of the Chlorinity/salinity Concept in Oceanography*. Elsevier, p. 227.

WARREN, B. A. (1973) Transpacific sections at latitudes 43° S and 28° S: the *Scorpio* Expedition, II. Deep water. *Deep-Sea Research*, **20**, 9–38.

WARREN, B. A. (1981) Deep circulation of the world ocean. Ch. 1, pp. 6–41, in *Evolution of Physical Oceanography*, B. A. WARREN and C. WUNSCH (Eds.), MIT Press.

WERTHEIM, G. K. (1954) Studies of the electrical potential between Key West, Florida and Havana, Cuba. *Transactions, American Geophysical Union*, **35**, 872–882.

WILLIAMS, A. J. (1975) Images of ocean microstructure. *Deep-Sea Research*, **22**, 811–829.

WILSON, B. W. (1960) Speed of sound in sea-water as a function of temperature, pressure and salinity. *Journal of the Acoustical Society of America*, **32**, 641–644.

WILSON, T. R. S. (1975) Salinity and the major elements in sea-water. Ch. 6, pp. 365–413 in *Chemical Oceanography*, Vol. 1, 2nd edn., J. P. RILEY and G. SKIRROW (Eds.), Academic Press.

WOOSTER, W. S. and J. L. REID (1963) Eastern boundary currents. Ch. 11, pp. 253–280, in *The Sea: Ideas and Observations*, Vol. 11, M. N. HILL (Ed.), Wiley-Interscience.

WORTHINGTON, L. V. (1954) Preliminary note on the time-scale in the North Atlantic circulation. *Deep-Sea Research*, **1**, 244–251.

WORTHINGTON, L. V. (1955) A new theory of Caribbean bottom water formation. *Deep-Sea Research*, **3**, 82–87.

WORTHINGTON, L. V. (1976) On the North Atlantic circulation. *Johns Hopkins Oceanographic Studies*, No. 6, p. 110.

WORTHINGTON, L. V. (1981) The water masses of the world ocean: some results of a fine-scale census. Ch. 2, pp. 42–69, in *Evolution of Physical Oceanography*, B. A. WARREN and C. WUNSCH, (Eds.), MIT Press.

WUNSCH, C. (1981) Low frequency variability in the sea. Ch. 11, pp. 342–375, in *Evolution of Physical Oceanography*, B. A. WARREN and C. WUNSCH (Eds.), MIT Press.

WÜST, G. (1935) Die Stratosphäre. *Wissenshaftliche Ergebnisse der Deutschen Atlantischen Expedition Meteor 1925–1927*, **6**, 1 Teil, 2 Lief. p. 180. (*The Stratosphere of the Atlantic Ocean*, W. J. EMERY (Ed.), 1978, Amerind, p. 112.).

WÜST, G. (1957) Stromgeschwindigkeiten und Strommengen in den Tiefen des Atlantischen Ozeans. *Wissenschaftliche Ergebnisse der Deutschen Atlantischen Expedition* Meteor 1925–1927, **6**, 261–420.

WÜST, G. (1961) On the vertical circulation of the Mediterranean Sea. *Journal of Geophysical Research*, **66**, 3261–3271.

WÜST, G. and A. DEFANT (1936) Atlas zur Schichtung und Zirkulation des Atlantischen Ozeans. *Wissenschaftliche Ergebnisse der Deutschen Atlantischen Expedition* Meteor 1925–1927, **6**, 103 plates.

WYRTKI, K. (1961) The thermohaline circulation in relation to the general circulation of the oceans. *Deep-Sea Research*, **8**, 39–64.

WYRTKI, K. (1965) The average annual heat balance of the North Pacific Ocean and its relation to the ocean circulation. *Journal of Geophysical Research*, **70**, 4547–4559.

WYRTKI, K. (1971) *Oceanographic Atlas of the International Indian Ocean Expedition*. National Science Foundation, Wash., DC, p. 531.

WYRTKI, K. (1973) Physical oceanography of the Indian Ocean. pp. 18–36 in *The Biology of the Indian Ocean*, Springer-Verlag.

WYRTKI, K. (1974) The dynamic topography of the Pacific Ocean and its fluctuations. *Hawaii Institute of Geophysics*, Report HIG–74–5, p. 19, 37 figs.

WYRTKI, (1975) El Niño – the dynamic response of the Pacific Ocean to atmospheric forcing. *Journal of Physical Oceanography*, **5**, 572–584.

WYRTKI, K. (1980) The Hawaii–Tahiti Shuttle Experiment. *Tropical Ocean–atmosphere Newsletter*, No. 4, Oct. 1980, pp. 7, 8. (Unpublished manuscript.)

WYRTKI, K., E. FIRING, D. HALPERN, R. KNOX, G. J. McNALLY, W. C. PATZERT, E. D. STROUP, B. A. TAFT and R. WILLIAMS (1981) The Hawaii to Tahiti Shuttle Experiment. *Science*, **211**, 22–28.

YOUNG, F. B., H. GERARD and W. JEVONS (1920) On electrical disturbances due to tides and waves. *Philosophical Magazine*, Series 6, **40**, 149–159.

Index